Photovoltaic Manufacturing

Scrivener Publishing
100 Cummings Center, Suite 541J
Beverly, MA 01915-6106

Publishers at Scrivener
Martin Scrivener (martin@scrivenerpublishing.com)
Phillip Carmical (pcarmical@scrivenerpublishing.com)

Photovoltaic Manufacturing

Etching, Texturing, and Cleaning

Edited by
Monika Freunek Müller

Scrivener Publishing

WILEY

This edition first published 2021 by John Wiley & Sons, Inc., 111 River Street, Hoboken, NJ 07030, USA and Scrivener Publishing LLC, 100 Cummings Center, Suite 541J, Beverly, MA 01915, USA

For more information about Scrivener publications please visit www.scrivenerpublishing.com.

Wiley Global Headquarters
111 River Street, Hoboken, NJ 07030, USA

For details of our global editorial offices, customer services, and more information about Wiley products visit us at www.wiley.com.

Library of Congress Cataloging-in-Publication Data

ISBN 978-1-119-24189-8

Cover image: Top figures: RENA Technologies GmbH – BatchTex machine for solar wafer texturing
Bottom row left: Technological Institute of Microelectronics (TiM), University of the Basque Country (UPV/EHU), Bilbao, Spain
Bottom row right: Fraunhofer Institute for Microstructure of Materials and Systems IMWS, Research Unit Center for Silicon Photovoltaics CSP, Halle (Saale), Germany
Cover design by Russell Richardson

Set in size of 11pt and Minion Pro by Manila Typesetting Company, Makati, Philippines

Printed in the USA

Contents

Preface

The last two decades were groundbreaking for photovoltaic (PV) technology. Countless researchers, engineers, technicians, politicians, and individuals all over the world contributed with their work and enthusiasm to the progress of this field. In this time, silicon PV cells increased their efficiency to 26.1% [1], being close to their theoretical limit for real cells of 29.8% [2]. PV technologies such as multijunction solar cells achieved a maximum of 39.2% efficiency in nonconcentrated applications [1], and new emerging technologies such as perovskites evolved. Figures 1 and 2 visualize the impressive progress in photovoltaics, depicting the best research cell efficiencies (Figure 1) and the champion module efficiencies (Figure 2). Both figures start with a few technologies, remarkable achievements, and, especially in the case of modules, a somewhat steady progress. The first cell type ever recorded in these data, an a:Si:H cell, evolved from 2.4% efficiency in 1976 to 14.1%. However, shortly around the year 2000, 10 new photovoltaic technologies evolved, increasing the record PV efficiency from 31.9% in 2000 by more than 50% to 47.1%. Not only were 40% of the technologies of today developed during the last two decades but also their efficiency trends are improving much steeper than ever before in the PV history.

The same is true for the record PV module efficiencies depicted in Figure 2. The efficiencies demonstrated with small-scale modules in the year 2000 increases with some module technologies doubling or even tripling their module record efficiencies within a decade and less. This is especially remarkable as some the technologies were scaled up to large module productions. Again, more than 40% of today's module technologies depicted in Figure 2 were not evolved at module or even research scale just 20 years ago.

The global solar module production reached the Gigawatt range soon after 2000 and ramped up to 165 GW in 2020 [3].

The use of PV technology has changed from test and research sites and first operational installations to being an essential part of national energy strategies worldwide. In many regions, PV is part of the landscape in both rural and urban areas. The levelized cost of energy (LCOE) is a common measure to

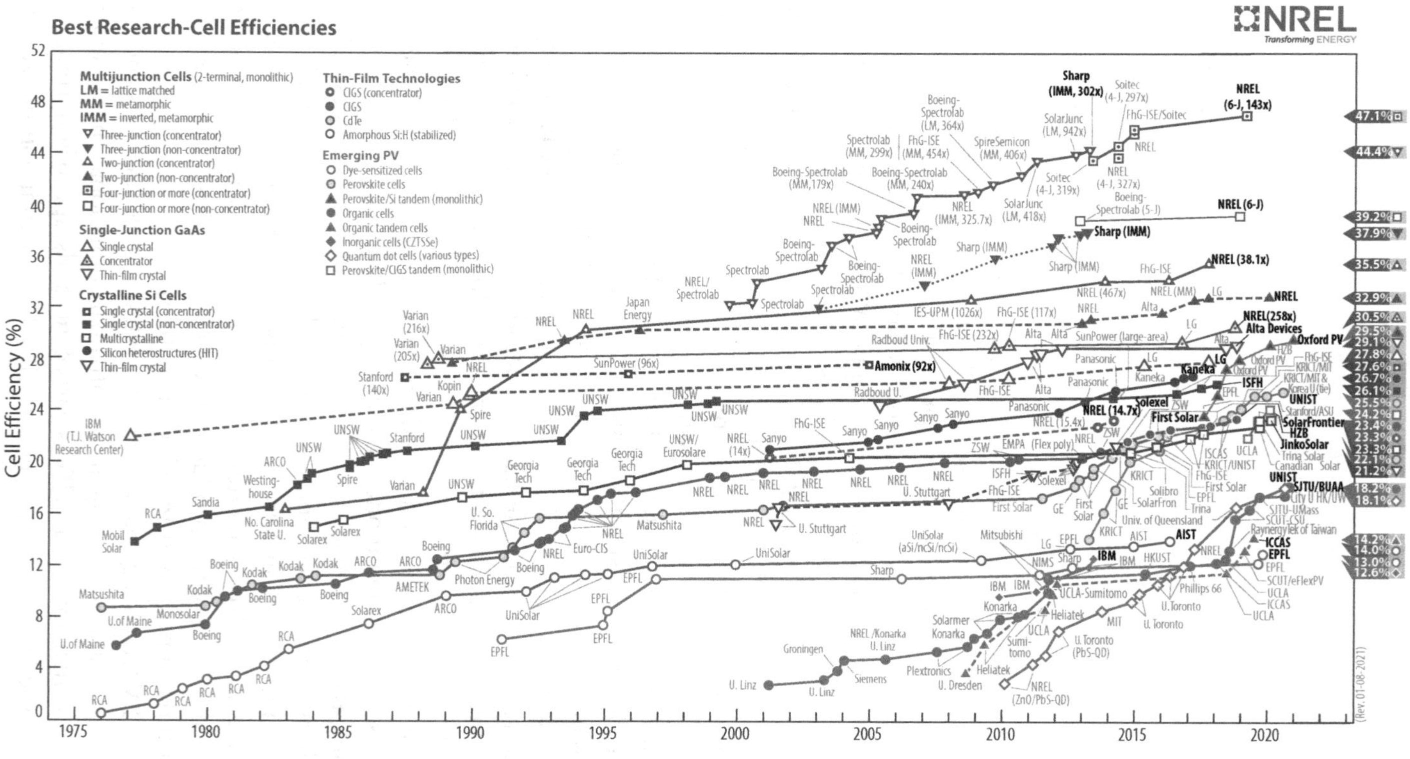

Figure 1 Conversion efficiencies of best research solar cells worldwide from 1976 through 2020 for various photovoltaic technologies. Efficiencies are determined by certified agencies/laboratories. Image by Nikos Kopidakis, National Renewable Energy Laboratory (NREL), Golden, CO, 2021, under public domain.

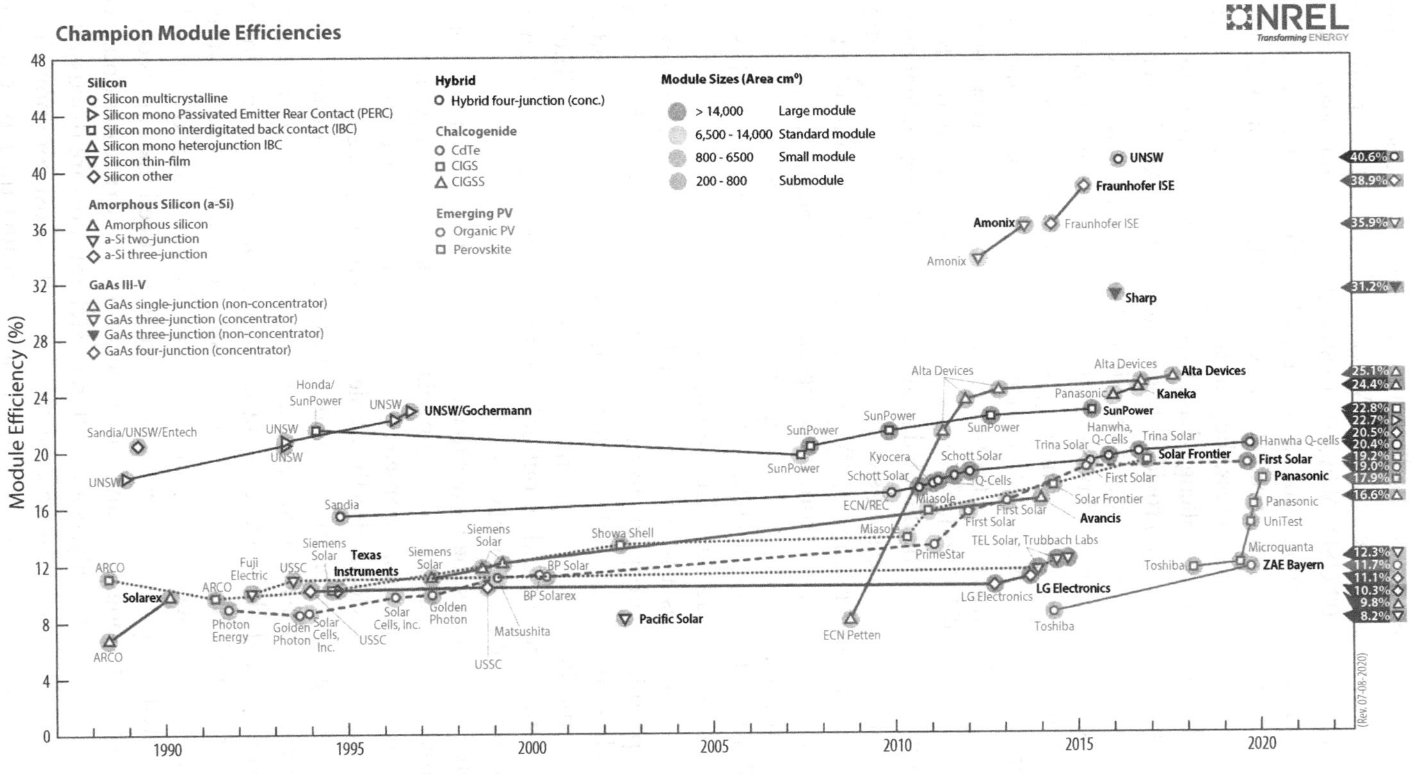

Figure 2 NREL chart of the highest confirmed conversion efficiencies for champion modules for a range of photovoltaic technologies, plotted from 1988 to the present. Image by Nikos Kopidakis, National Renewable Energy Laboratory (NREL), Golden, CO, 2021, under public domain.

compare the average net cost of different electric energy production technologies over their lifetime. The LCOE of PV per MWh has taken a breathtaking journey from about 400 USD/W in 2010 to 50 USD/MWh and lower globally with some projects reaching values as low as 23 USD/MWh [4, 5]. This trend is expected to continue with a predicted LCOE of 20 USD/MWh for 2030.

Despite this impressive progress, the processing of PV still is far from having reached its limits, and new challenges have to be addressed. Emerging developments, such as black silicon, provide a huge potential to make PV even more competitive in the field of energy conversion. Production efficiency requires a minimization of material and process losses, reproducible results, and economic scaling of the technology. The long-term nature of most PV applications and their large-scale implementation increases the importance of recycling. Ideally, this is included as part of the processing and manufacturing strategy.

The processing of PV today follows well-established standards, but as anyone involved knows, the detailed result will be highly dependent on the local machines and processing steps. Any difference in the settings might make a critical difference in the PV product performance and might distinguish the market leader from its competitors. This book introduces the readers to the theory and practical aspects of solar processing.

Metal-assisted chemical etching (MacEtch) for black silicon (b-Si) is expected to be the leading solar manufacturing technology in the future. Chapter 1 introduces this micro-/nanofabrication approach as one of the most promising prospects to further reduce the costs of photovoltaic devices while increasing their efficiency. The origin of MacEtch and the underlying mechanism are explained with a special focus on b-Si. The history, the state of the art, and an outlook toward the large-scale deployment in silicon photovoltaic industry are given.

Chapter 2 introduces the reader to alkaline texturing for the reduction of optical losses in monocrystalline silicon solar cells. The underlying process and the most important factors, parameters, and issues are explained. In addition, the texturing process is located in the whole manufacturing process of the solar cell, highlighting the importance of the previous steps for a high-quality result.

Chapter 3 provides a detailed introduction to advanced texturing with metal-assisted chemical etching in silicon solar wafers in general. The underlying electrochemical mechanisms are explained. Common methods, typical process steps, and structure characteristics obtained by metal-assisted chemical etching methods are introduced. Examples of the characteristics of topography and anti-reflection of the structures obtained using different metal catalysts and different etchant ratio are discussed.

Wet chemical cleaning of wafer surfaces and the most common cleaning technologies are outlined and discussed regarding their potential in the solar manufacturing process in Chapter 4. The reader is introduced to types and impact of contamination and to the concept of "contamination management." Examples for this innovative approach are given. The chapter closes with an economic perspective on the topic.

Reliable quality control, reproducibility, and the development of processing technologies all rely on analytics. Chapter 5 covers impurity analytics for the manufacturing of photovoltaic solar cells. With a special focus on the chemical analysis of silicon wafer surfaces, a detailed description of the analysis of trace metals is given. Current developments in analytical techniques for organic contamination are reviewed, and an overview on recent analytical techniques with application examples is provided.

This book is a comprehensive review on the most important steps in processing a high-quality solar wafer while keeping track of economic key values. The essential knowledge is explained by recognized experts in their field of endeavor. Outlooks on future topics are given, and recent challenges and innovations are presented. This book provides you with an efficient and solid start to this important field of photovoltaics. The technological limits of photovoltaic are still to be reached—may this compilation help you in exploring them.

The editor would like to thank Karen Reinhardt for the initial idea and the beginning of this project. I also thank sincerely all the authors for their willingness to share their expertise, their efforts to make their knowledge understandable for a larger audience, and for staying patient and focused during the publishing process of a book in the midst of a pandemic.

Monika Freunek
Lighthouse Science Consulting and Technologies, Canada
June 2021

References

1. F. Haase *et al.*, Laser contact openings for local poly-Si-metal contacts enabling 26.1%-efficient POLO-IBC solar cells, *Sol. En. Mat. Sol. Cells*, 186, November 2018, 184–193.
2. T. Tiedje *et al.*, Limiting Efficiency of Silicon Solar Cells, *IEEE Trans. El. Dev.*, 31(5), May 194, 711–716.
3. *IEA, https://www.iea.org/data-and-statistics/charts/solar-pv-module-manufacturing-and-demand-2014-2020, accessed 21-02-2021.*
4. *https://en.wikipedia.org/wiki/Cost_of_electricity_by_source, accessed 21-02-2021.*
5. *https://www.pv-magazine.com/2020/04/30/lcoe-from-large-scale-pv-fell-4-to-50-per-megawatt-hour-in-six-months/, accessed 21-02-2021.*

1

Metal-Assisted Chemical Etching of Silicon: Origin, Mechanism, and Black Silicon Solar Cell Applications

Chenliang Huo, Jiang Wang, Haoxin Fu and Kui-Qing Peng*

Beijing Key Laboratory of Energy Conversion and Storage Materials, Department of Physics, Beijing Normal University, Beijing, China

Abstract
Metal-assisted chemical etching (MacEtch) of silicon in hydrofluoric acid (HF) aqueous solutions is a widely used top-down approach for silicon micro/nanofabrication due to its cost-effectiveness, simplicity, versatility, and scalability. This method has recently emerged as a powerful surface micro/nanostructuring technique for low-cost and scalable production of black silicon (b-Si) with excellent light trapping properties, which might lead to both efficiency increase and cost reduction of solar cells. This review of MacEtch of silicon provides a critical description of its origin and the understanding of underlying mechanism highlights the story of MacEtch b-Si from initial discovery, through engineering improvements, toward the large-scale deployment in silicon photovoltaic industry.

Keywords: Metal-assisted chemical etching (MacEtch), solar cells, black silicon, photovoltaic

1.1 Introduction

Widespread deployment of solar photovoltaics (PVs) is critical to meeting the world's growing energy demand, tackling fossil fuel shortage and mitigating climate change in future, but solar PV remains uncompetitive

**Corresponding author*: kq_peng@bnu.edu.cn

Monika Freunek Müller (ed.) Photovoltaic Manufacturing: Etching, Texturing, and Cleaning, (1–42) © 2021 Scrivener Publishing LLC

(expensive and unreliable) relative to other technologies such as current fossil fuels-based electricity generation [1–3]. It is acknowledged that low-cost high-efficiency PV cells could capture more of the sun's energy and hence make sunlight-generated electricity economically competitive with fossil fuels-generated electricity [2–7]. Over the last decades, whereas crystalline silicon (c-Si) has been the most dominant semiconducting material for commercial PV cells due to its low-cost, earth-abundance and reliability, poor infrared absorption resulting from its indirect band gap, as well as relatively high reflectivity resulting from conventional surface texturing technologies greatly hurts the cell efficiency and poses a major challenge to the large-scale deployment of silicon PV modules [8, 9]. Hence, new ideas that strive for efficient utilization of much sunlight were under development [10–21]. One especially promising high-efficiency silicon solar PV technology is the black silicon (b-Si) solar cell, which is based on the c-Si wafer with micro/nanostructured surface that appears black to human's naked eye since it could capture sunlight across a broad range of wavelengths and angles of incidence very efficiently [2, 15–22]. In addition, excellent light trapping capability of b-Si allows reduction of wafer thickness and even no specific application of antireflection coating (ARC), thereby is an ideal material candidate for low-cost high-efficiency PV applications.

Low-cost and scalable production of b-Si is critical for industrial silicon solar cells but remains a significant challenge. Over the last decades, reactive ion etching [23–27], electrochemical/electroless etching, [28–30], femtosecond laser microstructing [15, 32–34], and metal-assisted chemical etching [16, 17, 35, 36] have been developed to texture silicon surfaces on the micro- and nano-scale for b-Si production. Currently, one of the biggest trends in b-Si manufacturing is the adoption of metal-assisted chemical etching technology as it has many advantages: it is low-cost, rapid, no requirement for expensive facilities, and scalable for industrial application. Recent advances in the scalable production of b-Si by metal-assisted chemical etching have pushed forward its practical applications in high efficiency silicon solar cells. The success of metal-assisted chemical etching b-Si technology is closely linked to its simplicity and well compatibility with existing industrial silicon solar cell production facilities. This review covers the history, mechanism, methods and recent achievements on b-Si by metal-assisted chemical etching for PV application. This article provides the reader with critical understanding of metal-assisted chemical etching, the b-Si fabrication processes, and the roadmap for large-scale deployment of b-Si solar cells.

1.2 History and Mechanism of Metal-Assisted Chemical Etching of Silicon

Metal-assisted chemical etching (MacEtch or MACE) of silicon generally refers to wet etching of silicon in the presence of noble metal particles or their film with openings, which are physically or chemically introduced onto silicon surface to enhance silicon etching at open-circuit potential (OCP) in oxidizing hydrofluoric acid (HF) aqueous solutions [35–40]. Within the last couple of years, MacEtch of silicon in HF aqueous solutions has aroused great interest and is presently one of the most popular top-down approach for micro/nanofabrication due to its cost-effectiveness, simplicity, versatility, and scalability [39–56]; it is also becoming increasingly important in various applications ranging from industrial solar cells, thermoelectric devices, batteries to medical biosensor and drug delivery [16, 17, 57–76]. Generally, for etching at OCP one can distinguish two mechanisms: electrochemical and chemical. The difference between both mechanisms is whether free holes are involved. The etching process at OCP involving free holes and depending on the redox potential generally is referred to as electroless etching. The MacEtch of silicon at OCP in oxidizing HF aqueous solutions is a noble metal catalyst-mediated electrochemical charge transfer process in which free holes are involved and spontaneously galvanic currents flow between local silicon anode and local metal cathode sites [37–40, 77, 78]. Therefore, the MacEtch of silicon is considered electrochemical reaction in nature. Besides MacEtch [38], the terms metal-assisted etching, metal-catalyzed chemical etching [35] or metal-catalyzed electroless etching (MCEE) [79–81] could be observed in literature. Among these terms, the acronym MCEE more clearly reveals the fundamental aspect of silicon etching electrochemistry in the presence of metal. Although we prefer MCEE or metal-catalyzed wet etching, considering the MacEtch has become common as Prof. Li suggested in the 2014 Spring Meeting of the Materials Research Society, we used it throughout this article.

1.2.1 History of Metal-Assisted Chemical Etching of Silicon

The discoveries and studies of metal-enhanced silicon etching actually began in the early 1990s [37, 82–90]. It is critically important to monitor metallic contamination on silicon wafer surfaces in order to achieve high-performance ultra-large-scale integration (ULSI) devices. Metallic contaminants such as Au, Cu, Ag, Pt, and Pd generally have fatal effects on device characteristics and must be suppressed to below 10^{10} atom/cm^2. Therefore, the behaviors of metallic contaminants, especially noble metal ions and their metallic particles on silicon wafer surfaces in wet chemical cleaning solutions have been extensively investigated to understand the underlying mechanism of metal deposition and then remove them from the silicon surface efficiently. Ohmi and other researchers found that noble metal cations such as cupric ions deposit on silicon surface in the metallic state through charge transfer from silicon to metal cations at the silicon/solution interface, simultaneously induces silicon oxidation nearby metallic deposits and then induces surface pits and microroughness in dilute HF (DHF) solution cleaning [37, 83–89]. The noble metal cations reduction takes places by either withdrawing conduction band electrons of silicon or injecting holes into silicon valence band holes. Moreover, Morinaga *et al.* found that silicon surface becomes rougher when an oxidizing DHF-H_2O_2 cleaning is used remove noble metallic particles [83]. Figures 1.1a–c show the scanning electron microscope (SEM) images of silicon surface with ultrafine Au particles after DHF-H_2O_2 cleaning. Porous silicon with small holes were clearly observed on the silicon surface. No obvious differences could be observed between the initial state and after DHF-H_2O_2 cleaning for silicon surface without particles. Morinaga *et al.* suggested that Au

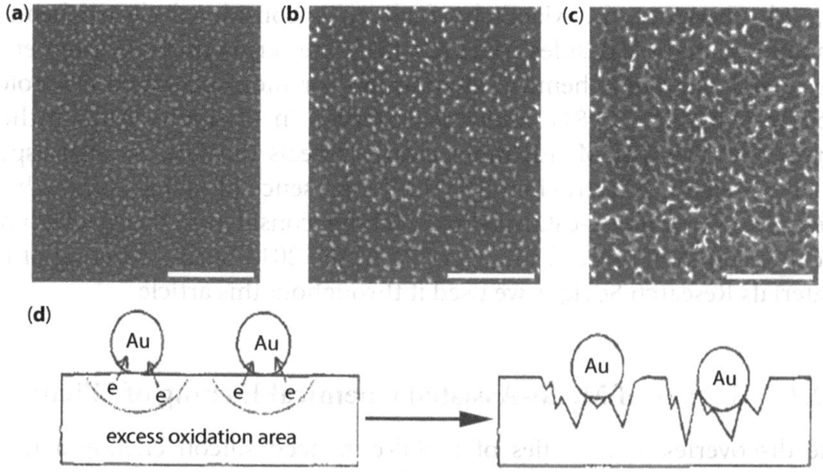

Figure 1.1 SEM images of the silicon wafers with Au particles after DHF-H_2O_2 cleaning. The silicon surface became rougher after cleaning. (a) initial. (b) 1 min. (c) 5 min. (d) Schematics of surface microroughness mechanism caused by Au particles with DHF-H_2. Reprinted with permission from H. Morinaga *et al.*, *J. Electrochem. Soc.* 142, 966 (1995). Copyright 1995 The Electrochemical Society.

features higher electronegativity than silicon and hence attracts electrons from silicon to induce silicon oxidation nearby it. Since the etch rate of SiO_2 by HF is always higher than its formation rate by H_2O_2, local excessive silicon oxidation and dissolution occur, resulting in silicon surface micro roughness or pitting due to irregular etching, as schematically shown in Figure 1.1d.

The pioneering fabrication of luminescent porous silicon by metal enhanced silicon etching was first demonstrated by Zhang and co-authors in 1993 [91]. They produced porous silicon layer on n-type silicon polished surface when the n-silicon back in contact with a noble gold metal was illuminated in a HF solution containing oxygen without an externally applied potential. In 1997, Dimova-Malinovska *et al.* reported the fabrication of porous silicon by etching an aluminium coated silicon substrate in HF-HNO_3 aqueous solution [82]. They claimed that the incubation time necessary for the porous silicon formation was dramatically decreased due to the presence of the Al film. In 1999, inspired by the discovery of Zhang *et al.*, Kelly and co-authors demonstrated that luminescent porous silicon layer can be made on p-type silicon surface in a similar way without illumination [92]. In their experimental setup as schematically shown in Figure 1.2a, a silicon wafer was short-circuited to an inert metal by evaporating inert Au/Cr film onto the back side of the wafer, thus a galvanic cell was formed. The resulting porous silicon layer with a thickness of 7 μm is shown in Figure 1.2b. They proposed that the porous silicon is achieved by galvanic etching due to the formation of metal/silicon galvanic cell in which the silicon acting as anode and metal as cathode in HF aqueous solution containing oxidants. The galvanic etch rate can be controlled by the metal/silicon area ratio and oxidant concentration of in HF solution.

Simultaneous platinum deposition and formation of a visible light-emitting porous silicon layer on silicon surface in fluoride solution containing platinum ions were reported by Gorostiza *et al.* in 1999 [93, 94]. They carried out electrochemical measurements and demonstrated that the hole injection from the platinum ions into the silicon results in the formation of porous silicon layer on silicon without requiring an electrochemical cell and voltage source. The SEM image depicted in Figure 1.2c shows the strongly etched silicon surface during Pt deposition in HF-K_2PtCl_6 solution. In 2000, Li and Bohn reported rapid preparation of light-emitting porous silicon by introducing metal nanoparticles (Au, Pt, or Au/Pd) to Si(100) surfaces prior to immersion in HF-H_2O_2 aqueous solution [38]. They demonstrated the metal coatings enhance silicon etching and result in a simple and effective way of producing porous silicon on the time scale of seconds. They termed the method "H_2O_2-metal-HF (HOME-HF) etching" in the main

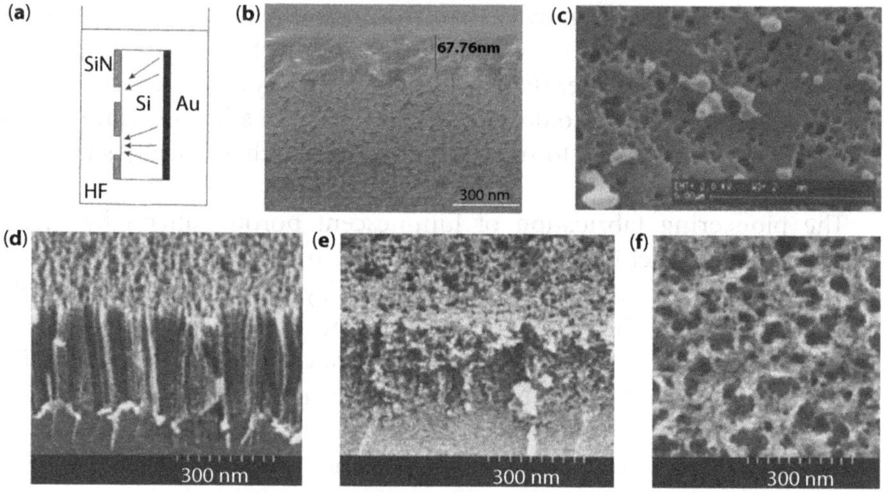

Figure 1.2 (a) Schematical view of the experimental setup with an Au electrode on the back side of the silicon wafer. (b) SEM image showing a cross-section of a porous Si layer formed using a gold-backed silicon wafer in an air-saturated HF solution. Reprinted with permission from C.M.A. Ashruf *et al., Sensor Actuat or A-Phys.* 74 118 (1999). Copyright 1999 Elsevier. (c) SEM image of porous silicon on silicon surface produced after Pt deposition in HF-K2PtCl6 solution. Reprinted with permission from P. Gorostiza *et al., J. Electrochem. Soc.* 144, 4119 (1997). Copyright 1997 The Electrochemical Society. (d, e) SEM images of Au-coated area on p+-Si(100) and off the Au-coated area on p+-Si(100) after etching in HF/H_2O_2 for 30s. (f) SEM image of Pt-coated area on p−-Si(100) after etching in HF/H_2O_2 for 30s. Reproduced with permission from Li XL *et al., Appl. Phys. Lett.* 77, 2572 (2000). Copyright 2000 American Institute of Physics.

text and "metal-assisted chemical etching (MacEtch)" in the title. Figures 1.2d, e show the SEM images of Au-coated p+ silicon etched in HF-EtOH-H_2O_2 aqueous solution for 30 seconds. Large interconnected pores are like the morphology observed on anodically etched porous silicon could be observed on the Au-coated areas. They found that strongly luminescent porous silicon can also be produced away from the metal-coated areas, implying lateral transport of charge carriers and chemical species during etching. Figure 1.2f shows the top-view SEM image of Pt-coated Si (100) surface after etching in HF/H_2O_2 for 30 s. A localized electrochemical process with the metal nanoparticle acting as a local cathode and silicon acting as an anode was proposed. Bohn and coauthors subsequently extended the MacEtch method to other semiconductors [95, 96].

In 1960, Turner has proposed that the wet etching of silicon in HF aqueous solution containing oxidant such as nitric acid (HNO_3) at OCP is an electrochemical process [97], in which silicon oxidation and dissolution takes place at local anode areas while the oxidizing agent is reduced at local cathode areas. Turner suggested that an etch pit will form at one site when it is anodic much more than it is cathodic while hillock will be produced at area that is cathodic more than it is anodic. However, the etching process is stochastic and non-preferential since any given area on silicon surface continually alternate between being anode and cathode. Inspired by the Turner's idea on the galvanic etching of silicon in HF solutions, Peng put forward and elaborate on a silicon micro/nanofabrication idea where [39, 98], locking micro/nanoscale anode or cathode sites leads to selective etching of silicon without mask in oxidizing HF aqueous solutions. They predicted that large-area silicon micro/nanostructures such as high-aspect ratio nanowires and nanoholes could be produced on silicon surface if something makes the idea come true. After many efforts, large-area aligned silicon nanowire (SiNW) arrays were firstly prepared by single-step silver-catalyzed etching of silicon wafer in HF-$AgNO_3$ aqueous solution in 2001, and they initially named the process self-assembling nanoelectrochemistry [39]. They ascribed the selective etching of silicon to the numerous microscopic localized silver/metal galvanic cells that are produced on silicon surface during etching in HF-$AgNO_3$ solution. Their first proposed etching process for SiNW formation in 2002 is incorrect, but subsequently were corrected based on compelling experimental results. Figure 1.3 shows the representative SEM images of the silicon substrates after etching in in HF-$AgNO_3$ aqueous solutions. Besides the silver dendritic film on the top of SiNW array, many silver nanoparticles can be observed at the interface between SiNW arrays and the intact Si substrate. The cross-sectional SEM observations confirmed that the silver particles induced selective etching of silicon, as shown in Figure 1.3 [99]. Aligned SiNWs also could be produced on silicon surface after selective etching in HF-$KAuCl_4$ aqueous solution [100, 101].

1.2.2 Mechanism of Metal-Assisted Chemical Etching of Silicon

So far, great efforts have been made to reveal the underlying mechanism of metal (e.g., Ag, Au, Pt, Pd, and Cu) assisted etching of silicon wafer in HF aqueous solutions containing strong oxidants (e.g., noble metal ions, nitrate, nitric acid, H_2O_2, and dissolved O_2) [38–126]. It has been revealed that the behavior of MacEtch varies with the etchant compositions and doping types of silicon, silicon at the metal-silicon interface undergone fast etching and the morphologies of resulting silicon micro/nanostructures depend on the shape or pattern of metal catalysts on silicon surfaces. Despite the fact that the MacEtch continues to dazzle us with its promising applications and unusual behaviors, the underlying physics and chemistry

Figure 1.3 (a) SEM image of aligned SiNW array prepared by one-step etching in HF-$AgNO_3$ solution. Reprinted with permission from K.Q. Peng *et al., Adv. Mater.* 14, 1164 (2002). Copyright 2002 Wiley-VCH. (b) SEM image showing aligned SiNW array on p-type (111) Si wafer etched in HF/$AgNO_3$ solution for 5 min. (c, d) SEM images showing aligned SiNW array on p-type (111) Si wafer etched in HF/$AgNO_3$ solution for 30 min. Reprinted with permission from K.Q. Peng *et al., Chem. Eur. J.* 12, 7942 (2006). Copyright 2006 Wiley-VCH.

of the etching system are still not well understood. For example, recent work clearly demonstrated that metal/silicon galvanic cell formation depends on the type and concentration of oxidants in HF aqueous solution [78]. Nonetheless, the mechanisms involving metal particles mediated charge transfer, metal particles catalyzed galvanic silicon etching, and the movement of metal particles into bulk silicon in unison have been widely accepted.

Since the galvanic cell-based silicon micro/nanofabrication idea came true [39], Peng and coauthors have systematically investigated the characteristics of MacEtch of silicon in HF aqueous solution containing oxidants [35, 40, 57, 77–81, 98–102]. They found that metal particles or metal film with openings induce fast dissolution of silicon underneath them and move into bulk silicon in the same direction during etching, and the motion direction of metal particles may suddenly change in unison due to some unknown reasons in some cases; the catalytic activity of the metal influences the oxidant reduction rate and thereby the etching rate. They suggested that the electrically coupled metal-silicon spontaneously constructs a microscopic short-circuited galvanic cell in which metal acts as local cathode for oxidant reduction while the silicon beneath metal acts as local anode and is subjected to oxidation and dissolution in HF solution [77], as shown in Figure 1.4a. For the sake of simplicity, the most widely used MacEtch system Ag-Si-HF-H_2O_2 is discussed here. The sustained silicon oxidation/dissolution and in unison movement of metal particles inward bulk silicon eventually leads to the formation of silicon micro/nanostructures such as nanowire and nanoholes (Figures 1.4b, c).

On the basis of experimental results [78], silicon dissolution during MacEtch of silicon in HF solutions consisting of divalent and tetravalent dissolution processes was proposed, and the two half-cell reactions that include cathodic and anodic reactions are expressed in the following equations (Equations 1.1–1.4). Note that the additional cathodic reaction expected to occur on metal surface or silicon surface is the reduction of noble metal ions since their metallic form may dissolve in the oxidizing HF solution, while the strong oxidizing solution environment does not favor the cathodic reduction of hydrogen ions. Clearly, besides the highly localized silicon dissolution at the metal/silicon interface, the overall electrochemical reactions in MacEtch of silicon are identical to those in electrochemical etching or stain etching of silicon in HF aqueous solution containing strong oxidants [27, 82, 128].

Cathodic reduction of oxidant at metal surface:

$$H_2O_2 + 2H^+ + 2e^- \rightarrow 2H_2O \tag{1.1}$$

Anodic oxidation of silicon in contact to metal:

$$Si + 2H_2O + 4h^+ \rightarrow SiO_2 + 4H^+ \tag{1.2a}$$

$$SiO_2 + 6HF \rightarrow H_2SiF_6 + 2H_2O \tag{1.2b}$$

$$Si + 6F^- + 2H^+ + 2h^+ \rightarrow SiF_6^{2-} + H_2 \tag{1.3}$$

Overall:

$$Si + 6HF + \frac{n}{2}H_2O_2 \rightarrow H_2SiF_6 + nH_2O + \left(\frac{4-n}{2}\right)H_2 \tag{1.4}$$

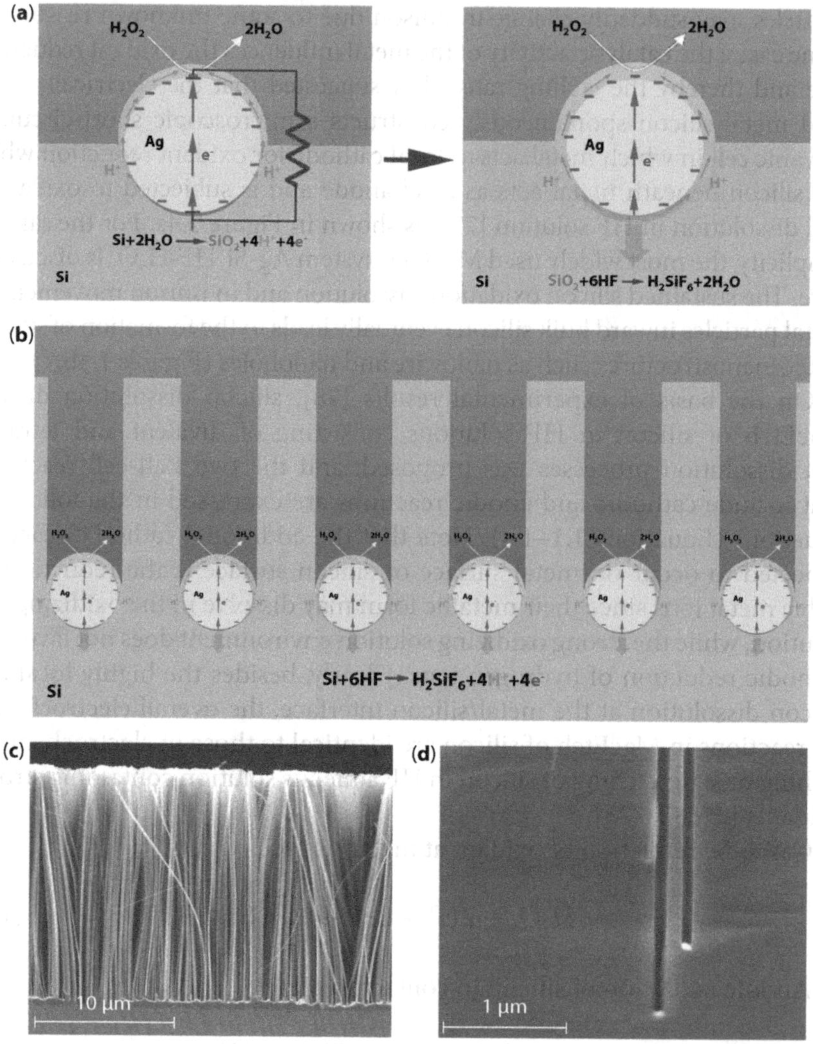

Figure 1.4 (a, b) Schematics of metal-assisted etching of silicon in HF-H_2O_2 aqueous solution. (c) Cross-sectional SEM image of aligned SiNW array produced on p-type 7–14 Ωcm Si (100) wafer by silver-assisted etching in aqueous HF-H_2O_2 aqueous solution for 30 min. (d) Cross-sectional SEM view of an etched Si(100) wafer, showing straight nanoholes normal to the silicon surface and Ag particles at the bottom of the nanoholes. Reprinted with permission from K.Q. Peng *et al., Adv. Funct. Mater.* 18, 3026 (2008). Copyright 2008 Wiley-VCH.

It is acknowledged that the MacEtch of silicon is an electrochemical process in which hole-transfer process is involved across the metal/silicon interface. Recently, Wang and coauthors experimentally and theoretically revealed that the band bending at the metal/silicon interface influences the hole transport and thereby the behavior of MacEtch of silicon at OCP [78]. Figures 1.5a, b illustrate the effect of H_2O_2 concentration on the band bending at Ag/Si interface for n- and p-type silicon with no surface states in HF aqueous solution. With the increase of H_2O_2 concentration, the excessive hole accumulation generates an additional bias at the Si/Ag interface and thereby change the interfacial band bending. At low H_2O_2 concentration, the rate of MacEtch for p-type Si is faster than that for n-type Si, while at high H_2O_2 concentration, the rate of MacEtch of silicon tends to be independent of doping types. Due to the excessive hole accumulation at Ag/Si interface area, a spontaneous electric field generate and induce silicon etching analogous to anodic etching of silicon at applied external

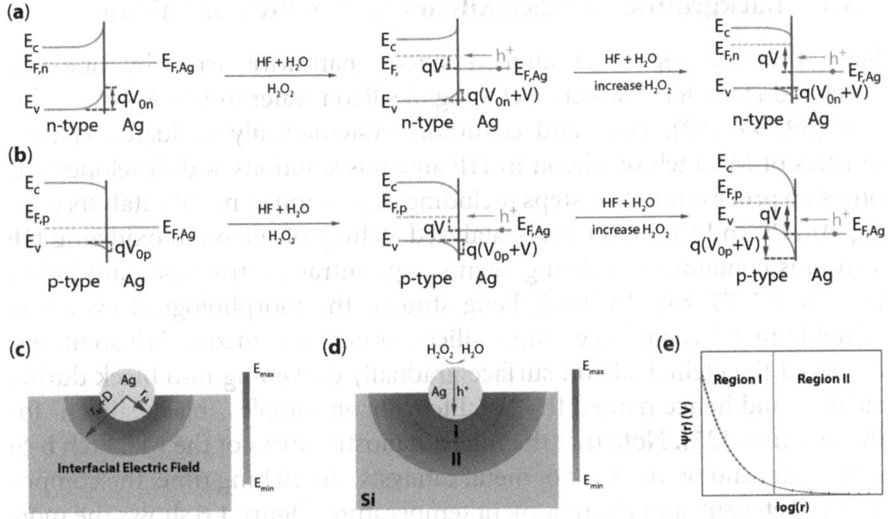

Figure 1.5 Energy band diagrams of energy band bending induced by H_2O_2 concentration change. (a) n-Si and Ag in contact. (b) p-Si and Ag in contact. (c) Schematic illustration of an embedded spherical silver particle in silicon and the induced interfacial electric field associated with Si/Ag contact before immersed into the aqueous HF-H_2O_2 solution. (d) The electric field distribution at Si/Ag interface region in aqueous HF- H_2O_2 solution. Region I: electropolishing region close to Ag particle, Region II: porous Si formation region. (e) The electrical potential $\Psi(r)$ as a function of log(r) (r is the distance from the centre of the Ag nanoparticle). Reprinted with permission from J. Wang *et al., Adv. Mater. Interfaces* 5, 1801132 (2018). Copyright 2018 Wiley-VCH.

bias (Figure 1.5d). The spontaneous interface electrical field could be modulated by the oxidant concentration. The interface electrical field distribute toward silicon around silver particle and could be divided into two regions according to the electrical potential: the electropolishing region with close to silver particle and the porous Si formation region away silver particle. Two-electron electrochemical oxidation of silicon (Equation 1.2a) occurs in the porous silicon formation region away from the silver particle, while four-electron electrochemical oxidation of silicon (Equation 1.2b) occurs at the electropolishing region close to silver particle (Figure 1.5e). This model naturally explains the porosification of silicon around metal particles with no need to consider secondary metal ions reduction and hole diffusion away from the metal/silicon interface area.

1.3 Fabrication and Optical Properties of Black Silicon by MacEtch of Silicon

1.3.1 Background of Black Silicon by MacEtch of Silicon

Since the first report of aligned silicon nanowire array by one-step silver-induced highly selective etching of silicon wafer in HF-AgNO3 solution [39, 99–102], Peng and coauthors systematically evaluated characteristics of MacEtch of silicon in HF aqueous solutions and developed the one-step process into two steps including deposition of metal catalysts (e.g., Ag, Au, Pt, and Cu) and metal-catalyzed etching of silicon in oxidizing HF solutions containing oxidizing agents (e.g., nitrate, nitric acid, and H_2O_2) [35, 40, 57, 77–81]. In 2003, Peng studied the morphological evolution of noble metal-coated crystalline silicon wafer in oxidizing HF solutions, observed the etched silicon surface gradually darkening into black during etching, and hence named the MacEtch silicon samples "black silicon" for the first time [35]. Note that the micro/nanostructures of the MacEtch b-Si surface depend on the kind of metal catalysts, the etching time, the composition of etchant, and the treatment temperature. Figure 1.6 shows the morphological evolution of a silver coated silicon wafer in HF aqueous solution containing oxidizing nitrate. With the elapse of etching time, silver particles gradually dug into bulk silicon and led to the formation of aligned SiNW array on the silicon surface [35]. Peng further suggested that a variety of silicon nanostructures with controlled morphologies can be produced with patterned metal catalysts on silicon surfaces [35]. SiNW-based b-Si and b-Si with various microstructures also could be produced by gold nanoparticles catalyzed etching of silicon in oxidizing HF solutions (Figure 1.6e), while

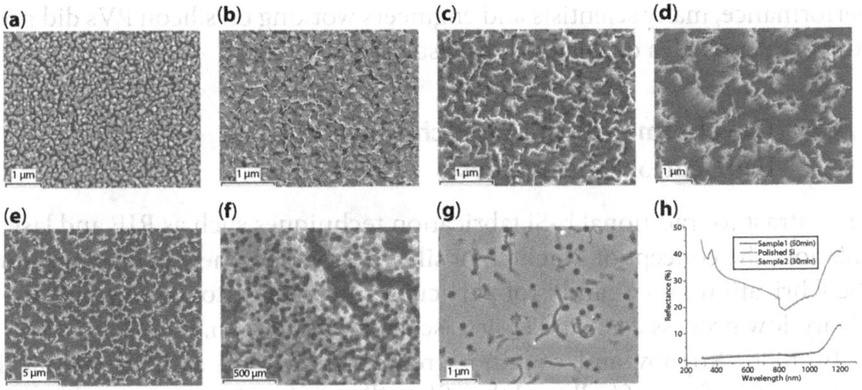

Figure 1.6 (a–d) SEM images showing the morphological evolution of a silicon wafer in HF/$Fe(NO_3)_3$ aqueous solution at t = 0, 2, 10, 50min. (e) SiNW array prepared by gold nanoparticles catalyzed etching in HF-$Fe(NO_3)_3$ aqueous solution. (f) Dense shallow pits prepared by copper nanoparticles catalyzed etching in HF-$Fe(NO_3)_3$ aqueous solution. (g) Nanoholes prepared by platinum nanoparticles catalyzed etching in HF-$Fe(NO_3)_3$ aqueous solution. (h) Reflectance (R) as a function of wavelength. Samples 1 and 2 were prepared by treating a Ag-covered silicon wafer in HF-$Fe(NO_3)_3$ aqueous solution for 50 and 30 min, respectively. Reprinted with permission from K.Q. Peng *et al.*, *Angew. Chem. Int. Ed.* 44, 2737 (2005). Copyright 2005 Wiley-VCH.

shallow pits (Figure 1.6f) are produced by copper nanoparticles catalyzed chemical etching since copper is not stable in the oxidizing environment and would dissolve into the solution [35]. Dispersive nanoholes are generally produced in case of dispersive platinum nanoparticles catalyzed etching (Figure 1.6g). All the metal catalysts on silicon wafers are prepared by low-cost and scalable wet electroless deposition methods for practical applications [35]. For example, the silicon wafers were immersed in HF aqueous solution containing $AgNO_3$ for Ag catalyst deposition and immersed in HF aqueous solution containing $KAuCl_4$ or $HAuCl_4$ for gold catalyst deposition. The MacEtch b-Si suppresses light reflection less than 1.46% or more less over a wide spectral bandwidth, thereby represents an ideal sunlight absorbing material for high-efficiency silicon solar cells (Figure 1.6h). Peng ascribed the remarkable light trapping properties under the solar spectrum to the following possible reasons: 1) an ultrahigh surface area; 2) the subwavelength-structured (SWS) surface of the surface microstructures; 3) gradient change of the refractive index with etching depth. In contrast to RIE and laser ablation for b-Si fabrication, this all-wet MacEtch b-Si technology is quick, low cost, scalable, and well compatible with industrial silicon texturing processes for solar cells. Unfortunately, at that time, considering the notoriously detrimental effects of noble metals on device performance, many scientists and engineers working on silicon PVs did not support Peng's idea on MacEtch b-Si solar cell concepts.

1.3.2 Developments on MacEtch Black Silicon Fabrication Techniques

In contrast to traditional b-Si fabrication techniques such as RIE and laser ablation, the concept of MacEtch of silicon provides a new perspective for the fabrication of b-Si and is of particular interest due to its inherent simplicity, low process cost, and easy to scalable production.

In 2006, Koynov and coauthors reported the gold-assisted chemical etching of surfaces of bulk and thin film silicon of various structural forms and doping in HF-H_2O_2 aqueous solutions for the fabrication of b-Si surfaces [130], which result in dramatic suppression of reflectivity in a broad spectral range. The discontinuous gold particle layer was deposited by thermal evaporation at a pressure of 10^{-5}–10^{-6} mbar rather than simple wet chemistry methods. Figure 1.7a shows the photograph of b-Si surfaces on different silicon surfaces and their optical properties. All the surfaces of the b-Si samples appear similarly black even under non-normal incidence. The optical reflection measurement reveals the surface optical reflectivity can be reduced to as little as 2%–5% in the high light absorption ranges for all three samples. Figure 1.7b show the internal absorption spectra of the black etched crystalline Si and thin film a-Si:H. It can be observed that the amount of absorbed light is significantly increased due to the suppressed reflection after the nanoscale texturing of silicon surfaces with MacEtch treatment. As a faster and less expensive alternative, in 2009, Branz *et al.* described a one-step MacEtch technique that use mixed $HAuCl_4$, HF, and H_2O_2 aqueous solution to produce a graded-density b-Si surface layer of a network of nanopores [103]. Figure 1.7c shows the cross-sectional SEM images of the b-Si samples and dense nanopores could be seen near the surface. This nanopore network grades the near-surface optical constants and reduces optical reflectance to below 2% throughout the usable solar spectrum (350 to 1,000 nm). Subsequently, similar single-step MacEtch techniques that use HF and H_2O_2 aqueous solutions containing low-concentration noble metal ions such as Ag^+ and Cu^{2+}, could be developed to produce micro/nanostructured b-Si surfaces. In 2014, inspired by metal corrosion in air, Hu *et al.* demonstrate that MacEtch of silicon can be performed simply in aerated HF/H_2O vapor for facile fabrication of b-Si [81].

Copper-induced silicon surface pitting and porousing have been investigated since 1990s [37, 83–89]. For example, Peng *et al.* reported the formation of nanostructured silicon surface by copper-catalyzed etching of silicon

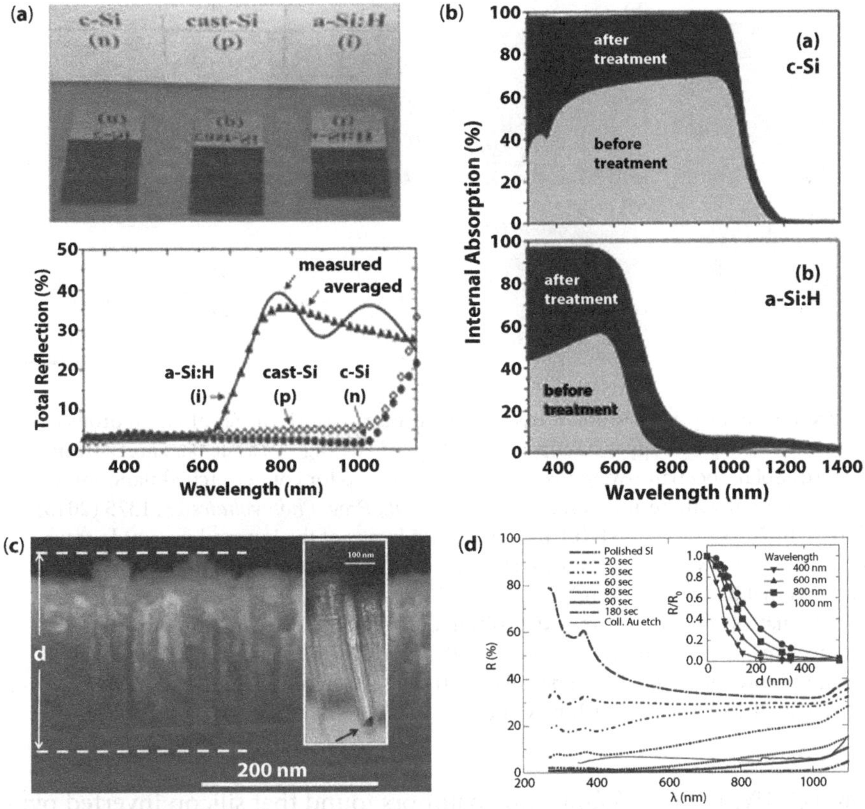

Figure 1.7 (a) Photograph of gold-catalyzed etched b-Si samples of different crystal structures and dopings and total hemispherical reflection spectra of the treated areas of these samples. (b) Internal absorption spectra of the black etched crystalline silicon and thin film amorphous Si:H in comparison with the corresponding absorption of the untreated samples. Reprinted with permission from K. Svetoslav *et al.*, *Appl. Phys. Lett.* 88, 203107 (2006). Copyright 2006 American Institute of Physics. (c) Cross-sectional SEM image of the Si (100) surface etched in HF-H_2O_2-$HAuCl_4$ solution for 80s. Inset to (c) is dark-field TEM image of a nanopore and Au particle (indicated by arrow) produced by HF-H_2O_2-$HAuCl_4$ solution etching. (d) Reflectance spectra taken after etching in HF-H_2O_2-$HAuCl_4$ solution for different etching times as indicated. Reprinted with permission from H.M. Branza *et al.*, *Appl. Phys. Lett.* 94, 231121 (2009). Copyright 2009 American Institute of Physics.

while pointed out that copper nanoparticles in HF aqueous solution containing strong oxidizing agents such as Fe^{3+} and H_2O_2 are instable and dissolve into the etching solution as soluble Cu^{2+} ions [35, 40]. In 2014, Toor *et al.* prepared nanoporous nanostructured b-Si surfaces by cheap copper-assisted chemical etching of silicon (Figure 1.8a) [129]. They demonstrated that the solar-spectrum-weighted average reflection of the copper-etched b-Si is as

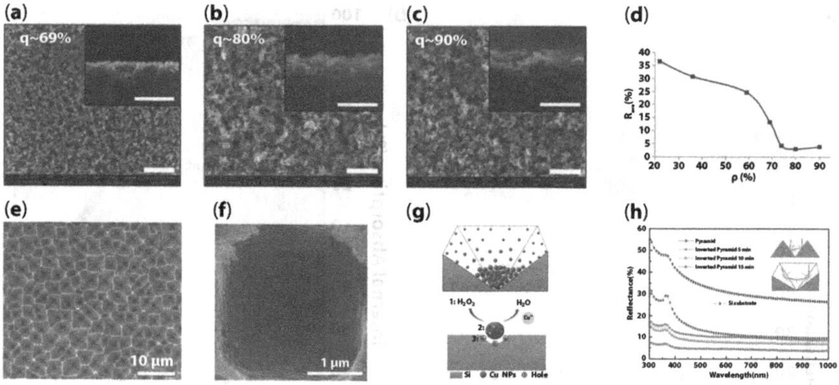

Figure 1.8 (a–c) Top-view and cross-sectional (insets) SEM images of nanoporous b-Si as a function of ρ by copper nanoparticle catalyzed etching. All scale bars are 500 nm long. (d) Solar-spectrum-weighted reflectance versus ρ for copper-etched nanoporous b-Si. Reprinted with permission from F. Toor *et al., Prog. Photovoltaics* 23, 1375 (2015). Copyright 2015 Wiley-VCH. (e) Top-view SEM image of the inverted pyramid arrays for 15 min processing in $Cu(NO_3)_2/HF/H_2O_2/H_2O$ solution. (f) SEM top-view image of an individual inverted pyramid for 1 min etching in $Cu(NO_3)_2/HF/H_2O_2/H_2O$ solution. (g) Schematics of the anisotropic deposition of Cu NPs, H_2O_2 reduction process and the mechanism for the holes injection. (h) Reflectance spectra of upright and inverted pyramid arrays. Reprinted with permission from Y. Wang *et al., Sci. Rep.* 5, 10843 (2015). Copyright 2015 Nature Publishing Group.

low as 3.1%. In 2015, Wang and coauthors found that silicon inverted pyramids could be produced on silicon surfaces by single-step copper nanoparticle assisted etching in a mixed $Cu(NO_3)_2$, HF, H_2O_2 aqueous solutions at 50°C [130]. Figures 1.8e, f show the large-area random inverted pyramid arrays produced on Si(100) surface. The length of the inverted pyramids' bottom side is about 2–6 µm and the depth is in the range of 1–5 µm. The copper assisted anisotropic etching of silicon is ascribed to the galvanic etching like the well-known metal-assisted chemical etching method for silicon nanowire fabrication. Meanwhile, they suggested that the anisotropic deposition of Cu nanoparticles during etching is the key step in the formation of inverted pyramids without any mask (Figure 1.8g), and the generation and dissolution of Cu nanoparticles should keep in balance during the etching to produce micro-structured inverted pyramids, which are affected by the concentration of the etchant, the doping type and the doping level of the silicon substrates. The reflectance spectra of the inverted pyramids are shown in Figure 1.8h. The optical reflectivity of the copper-etched inverted pyramids is reduced to ~4.4% over the wavelength range from 300 to 1,000 nm, much lower than that of the upright pyramids by conventional alkaline etching.

In summary, various MacEtch b-Si techniques have been developed [38–125, 128–132]. Having the low-cost and scalable production of b-Si in mind, all-wet silver and copper catalyzed etching of silicon are especially attractive for PV application.

1.4 Photovoltaic Solar Cell Applications of MacEtch Black Silicon

Due to its micro/nanoscale surface textures favoring broadband light absorption, MacEtch b-Si solar cells are expected to convert sunlight into electricity cost-efficiently and thereby boost solar power generation globally. Scientists and engineers around the world have continued to develop the MacEtch b-Si solar cell manufacturing techniques that could help make it more economically attractive.

1.4.1 Silicon Nanowire/Nanohole-Based MacEtch Black Silicon for Photovoltaic Solar Cells

Nanowire and nanohole-based b-Si has received intense attention as sunlight absorbers since it efficiently eliminates front-surface reflection with no need of conventional ARCs [16, 17]. Peng *et al.* demonstrated that MacEtch b-Si samples such as SiNW array drastically suppresses light reflection over a wide solar spectral bandwidth and suggested the high light absorption is due to the MacEtch b-Si surfaces with subwavelength microstructures [16, 35]. Then, the light trapping properties of silicon nanostructures with various shapes such as wire and hole have been investigated theoretically and experimentally [133–154]. It is believed that the remarkable light trapping capabilities of nanostructured silicon surfaces are due to the photonic crystal enhancement and long path length of incident light. Experimental and theoretical results revealed that silicon wire array with less silicon material achieves the same amount of light absorption as in conventional thick c-Si wafer [133, 143].

The MacEtch b-Si samples with remarkable light trapping properties are expected to convert a significant amount of useable sunlight into electricity at low cost [35, 36, 103, 128–132, 150–161, 163–187]. Peng and coworkers tried to prepare SiNW-based b-Si solar cells by silver-assisted chemical etching of single-crystalline silicon (sc-Si) and multicrystalline silicon (mc-Si) wafers and traditional silicon PV manufacturing technique for the first time [16, 98]. Figures 1.9a, b show the SEM images of SiNWs on sc-Si and mc-Si wafers. They fabricated the n-emitter by thermally diffusing

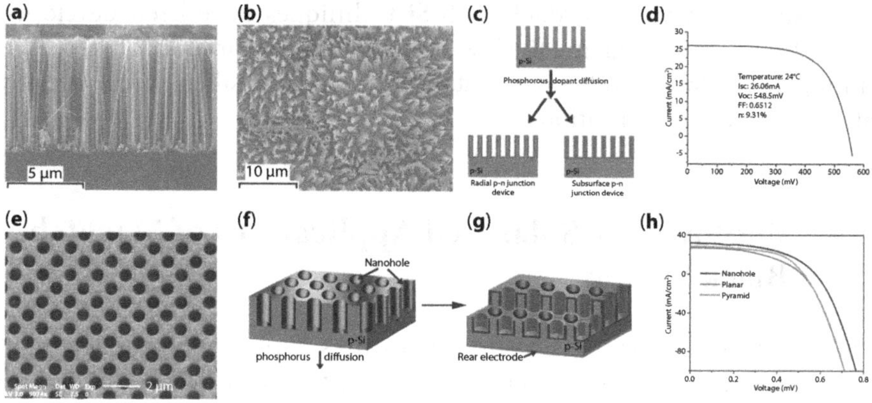

Figure 1.9 (a) SEM image of SiNW-based single-crystalline b-Si by silver-catalyzed etching. (b) SEM image of SiNW-based multi-crystalline b-Si by silver-catalyzed etching. (c) Schematic illustration of the fabrication of functional p-n junctions for SiNW-based b-Si solar cells. (d) I–V curve of SiNW-based monocrystalline b-Si solar cell. *Reprinted* with permission from K.Q. Peng *et al., Small* 1, 1062 (2005). Copyright 2005 Wiley-VCH. (e) Top-view SEM image of silicon nanoholes on c-Si(100) wafer surface by silver-catalyzed etching. (f) Schematic illustration of the fabrication of nanohole-based b-Si solar cell with radial p-n junctions via thermal phosphorus dopant diffusion. (g) I-V curve for solar cells with different geometries fabricated under identical conditions. Reprinted with permission from K.Q. Peng *et al., J. Am. Chem. Soc.* 132, 6872 (2010). Copyright 2010 American Chemical Society.

phosphorus from a $POCl_3$ source (Figure 1.9c). Front grid Ti/Pd/Ag and rear aluminum electrodes were prepared by thermal evaporation. No ARC and surface passivation layer were used for the SiNW-based b-Si solar cell. However, the conversion efficiency of the cell is not as high as expected from the excellent light absorption property. The efficiencies of SiNW-based b-Si solar cells on single crystalline silicon and multi crystalline silicon are 9.31% and 4,37%, respectively. Peng suggested that nanohole geometry exhibits superior mechanical robustness and optical absorption ability in comparison with free-standing fragile nanowire counterparts. Peng *et al.* further fabricated silicon nanohole array (Figure 1.9e) on p-type Si(100) wafer by silver-assisted chemical etching and deep ultraviolet lithography, and examined silicon nanoholes as potential light absorber candidate for high efficiency solar cell [17]. The nanohole solar cell exhibited a power conversion efficiency of 9.51% under 1 sun AM1.5G illumination (Figure 1.9g). The nanowire- and nanohole-based b-Si solar cells exhibited lower energy conversion efficiencies than conventional silicon solar cells due to the enhanced surface/interfacial recombination associated with high-surface-area nanostructures and poor device manufacturing conditions. They suggested that the structural optimization, suitable surface passivation, and improved cell manufacturing condition should enhance the performance of the nanowire and nanohole-based b-Si solar cells.

To reduce the quantity of used silicon materials, Sivakov *et al.* demonstrated a 2.5- to 3-μm-thick multicrystalline SiNW-based b-Si solar cell on inexpensive glass substrates [59]. The solar cell with axial p_+-n-n^+ junctions exhibited an energy efficiency of 4.4% under AM 1.5G illumination. In 2008, Yang group reported core-shell SiNW-based b-Si solar cell by coating p-type amorphous silicon onto n-type nanowires surfaces [60]. The cell showed efficiency of about 0.5%, and the extremely low efficiency was attributed to the high interfacial recombination and high series resistance of MacEtch nanowires. In 2015, Shen and coauthors employed a one-step MacEtch technique to prepare large-area silicon nanostructures with smoother surface morphology and a 17.63% efficiency of the MacEtch mc-Si solar cells surpassed that (17.45%) of traditional-micro-textured solar cells was demonstrated [157]. The key point lies in the electrical loss reduction by removing porous layer and employing shorter one-step-MacEtch-smoothened b-Si surface. It is known that spectral response of solar cells determines the output performance. In 2016, Huang *et al.* suggested that the passivated front together with the passivated rear surface of b-Si provide the complementary spectral responses in both the short-wavelength and long-wavelength regions. They fabricated b-Si samples with complex surface micro/nanostructures by anisotropic alkaline etching and subsequent one-step MacEtch of c-Si wafers. Simultaneous stack SiO_2/SiN_x passivation for the front and rear surface of b-Si was performed. The as-prepared b-Si solar cell possessed excellent broadband spectral response and exhibited a conversion efficiency of about 20.0% [158].

1.4.2 Alkaline Treatment Modified MacEtch Black Silicon for Photovoltaic Solar Cells

Typically, nanowire or nanoholes induced during MacEtch of silicon is accompanied by the simultaneous formation of microporous layer, which is very helpful to reduce surface reflectance, but impedes the formation of p-n junction while results in increased surface/interface recombination issue, thereby leading to solar cells with poor performances.

In 2006, Tsujino and coauthors reported the texturization of cast multicrystalline silicon wafers by platinum or silver particles assisted chemical etching in HF solution containing oxidant such as H_2O_2 [59]. Figure 1.10a shows the SEM image of a multicrystalline silicon surface covered with a

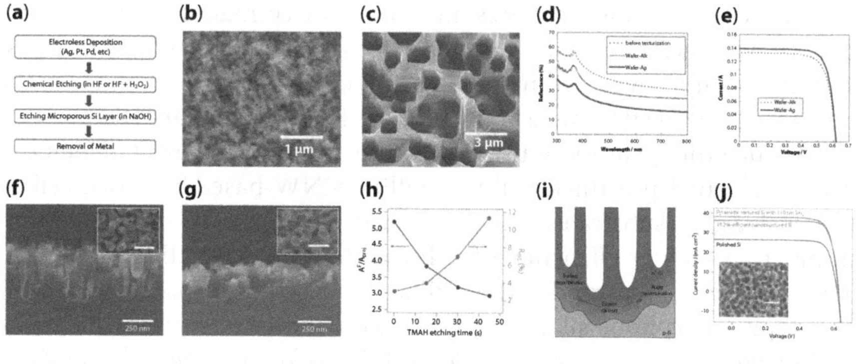

Figure 1.10 (a) Process sequence for texturization of mc-Si wafers by alkaline treatment modified MacEtch. (b) SEM image of microporous porous layer on etched mc-Si wafer surface. (c) SEM images of the mc-Si wafer surface after silver-catalyzed etching and NaOH etching. (d) Reflectance spectra of mc-Si wafers before texturization and those of MacEtch b-Si and alkaline etching. (e) I-V characteristic of solar cells made from alkaline treatment modified MacEtch and conventional alkaline etching. Reprinted with permission from K. Tsujino *et al.*, *Sol. Energy Mater. Sol. Cells* 90, 100 (2006). Copyright 2006 Elsevier. (f) Cross-sectional SEM images of MacEtch silicon nanostructure. (g) Cross-sectional SEM image of alkaline treatment modified MacEtch silicon nanostructure. (h) Measured surface area enhancement ratio and the solar spectrum averaged reflectance of silicon nanostructures as a function of TMAH etching time. (i) Photo-generated carriers (blue dots) in the n+-Si nanostructures (red regions in the FIG.) dissipate by Auger and surface recombination. (j) I-V curves of 18.2% efficient b-Si, polished silicon and pyramid-textured silicon with an SiNx antireflection coating. Reprinted with permission from J. Oh *et al.*, *Nat. Nanotechnol.* 7, 743 (2012). Copyright 2012 Nature Publishing Group.

microporous layer after silver-catalyzed etching in mixed 10% HF: 30% H_2O_2 (10:1) solution for 5 min. Considerable reduction of surface reflectance was achieved after silver-catalyzed etching. They found that the solar cells made from the MacEtch b-Si with microporous silicon layer did not show good performance as expected because the formation of p-n junction through the microporous silicon layer is rather difficult. Hence, they removed the microporous silicon layer by 1% sodium hydroxide (NaOH) etching for 18 min. Subsequently, the wafer surface is treated in 30% HNO_3 for 30 min to remove the metal particles on the surface. Figure 1.10b shows the SEM image of alkaline treatment modified MacEtch b-Si. It can be clearly seen that the alkaline modified MacEtch b-Si surface exposes crystalline facets of about 1 μm and consists of micro/nanoscale holes. At the bottom of holes on the (100) planes, inverted pyramids could be clearly observed (Figure 1.10c). The surface textures were considerably affected by the kind of metal catalysts, the composition of the HF/H_2O_2 solution, the etching time, and the conditions of the alkaline treatment. The averaged surface reflectance of MacEtch b-Si after anisotropic wet etching in NaOH aqueous solution is about 22% over the wavelength range from 300 to 800nm, much lower than that of multicrystalline silicon wafers textured by conventional alkaline etching, as shown in Figure 1.10d. Traditional thermally diffused-junction silicon solar cell processing technique was adopted. An n-layer emitter of 60 Ω/sq was prepared on the top of the NaOH treatment modified MacEtch b-Si surface by thermally $POCl_3$ diffusion. After chemical cleaning, a silicon nitride layer was used as ARC by plasma-enhanced chemical vapor deposition, and front silver and back aluminum electrodes were applied by the screen-printing method. Tsujino *et al.* demonstrated that the averaged performance of the solar cells fabricated from NaOH treatment modified MacEtch b-Si is better than those of the cells fabricated from conventional alkaline textured silicon wafers. This can be clearly seen from the measured current-voltage (I–V) characteristics shown in Figure 1.10e. In addition to the slight increase of IQE and efficiency, the NaOH treatment modified MacEtch b-Si solar cell showed slight increases in the filling factor (FF) and open circuit voltage (Voc). This work paved the way for scalable production of efficient MacEtch b-Si solar cells. In 2014, Ye and coauthors repeated Tsujino's work and fabricated pseudo-pyramid textured multi-wire slurry sawing (MWSS) multi-crystalline silicon solar cells with efficiency 18.45% by using silver-catalyzed etching and post alkaline etching on an industrial production line [159]. In 2015, Lin and coauthors prepared nanostructured mc-Si with controlled geometry and surface area by Ag-catalyzed etching and subsequent NaOH treatment. They demonstrated a nanostructured mc-Si solar cells with a certified efficiency of 17.75% on large size (156 × 156 mm^2), which is ~0.3% higher than its acid textured counterparts [160].

The enhanced surface/interface recombination in the MacEtch b-Si has severely limited the photo-generated carrier collection efficiency and hindered its use in PVs. The carefully designed surface doping and passivation schemes are highly expected to pave the way for the application of b-Si in existing and new high-efficiency solar cell concepts. In 2012, the researcher of the National Renewable Energy Laboratory demonstrated the photo-generated carrier transport/collection efficiency in MacEtch b-Si solar cells is sensitive to the p-n junction formation conditions and front surface passivation [22]. In addition to the surface recombination associated with high surface area, Auger recombination caused by excessive doping related to p-n junction formation through the high surface area of b-Si limits the photo-generated carrier collection efficiency. In addition, 18.2% efficient MacEtch b-Si solar cell without additional ARC can be reached by reducing surface area and simultaneously controlling thermal diffusion doping conditions. One key point in the fabrication of efficient b-Si solar cell is that b-Si samples prepared by silver-catalyzed etching of float-zone grown Si(100) wafers in HF-H_2O_2 aqueous solution underwent anisotropic wet etching in tetramethylammonium hydroxide (TMAH) solution. Figures 1.10f, g compare the cross-sectional SEM images of the b-Si solar cells without and with TMAH etch treatment. It can be clearly seen that TMAH treatment widens the pore diameters and simultaneously reduces the pore depths. As shown in Figure 1.10h, the anisotropic TMAH etching considerably decreases the front surface area of b-Si while enhances the optical reflectance. The b-Si solar cells are fabricated by conventional diffused-junction solar cell manufacturing technique using liquid $POCl_3$ as phosphorus diffusion source, and surface passivated with SiO_2 passivation film formed by thermal oxidation. Figure 1.10j shows the I-V curve of the best b-Si solar cell with 18.2% efficiency, 36.45 mA/cm^2 short-circuit current, 0.628 Voc, and 79.6% fill factor (FF) under AM 1.5G illumination. For this best solar cell, a 60-s etch in 1 vol% TMAH solution at room temperature was used after n^+-emitter formation to reduce surface phosphorus doping concentration and surface area, thereby both Auger recombination and surface recombination are simultaneously reduced. The SEM image shown in the inset in Figure 1.10j revealed the nanoscale inverted pyramids produced on the surface of b-Si solar cell after anisotropic TMAH etching. In 2018, Xu and coauthors prepared nanoscale inverted pyramids on c-Si wafers by Ag-catalyzed etching and alkaline etching. Through optimizing the cell design, 20.5% efficient b-Si solar cells with nanoscale inverted pyramid texture were achieved [161].

1.4.3 MacEtch Black Silicon for Diamond Sawed mc-Si Photovoltaic Solar Cells

Although silicon wafers for the PV industry are traditionally sliced using a MWSS technique in the last decades, diamond wire sawing (DWS) has emerged as the mainstream technique for wafer slicing due to its high productivity, wafer cost reduction, and environmentally friendly sawing process [162]. Compared to traditional MWSS process, DWS process considerably reduces the saw damage depth typically by 20%–50% and thereby less saw damage on wafer surfaces, which is critical when moving toward thinner wafers, but poses major challenge in texturing mc-Si with the standard etching processes. For example, 0.3% efficiency loss is often generated when DWS mc-Si wafers are processed with standard acidic HNO_3-HF etching process. Therefore, new surface texture technologies suitable for DWS wafers are highly expected.

Unlike traditional wet etching processes, MacEtch technique does not require any saw damage or defects to produce surface texture with excellent light trapping properties, thereby making it extremely useful for DWS wafers. In 2015, Cao and coauthors extended MacEtch and Tsujino's post alkaline treatment strategy to DWS mc-Si wafers. Both MWSS and DWS mc-Si wafers underwent the same normal acid etching and MacEtch to produce micro- and nanoscale textures [163]. Figure 1.11 shows the SEM

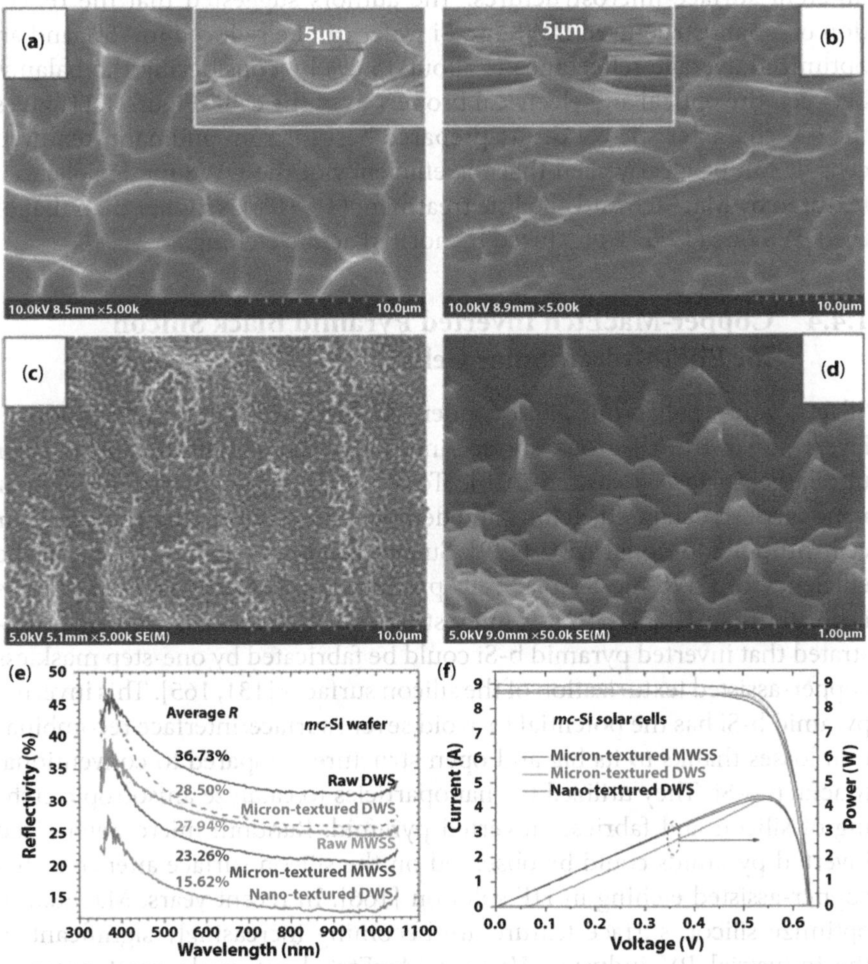

Figure 1.11 (a) SEM image of a conventional acid-textured MWSS mc-Si wafer. (b) SEM image of a conventional acid-textured DWS mc-Si wafer. (c, d) SEM images of an alkaline treatment modified MacEtch-textured DWS mc-Si wafer. (c) Reflectivity of MWSS and DWS mc-Si wafers with different microstructures. (d) I-V characteristics of mc-Si solar cells with different surface textures under AM 1.5G illumination. The measurement area is 15.6×15.6 cm^2. Reprinted with permission from F. Cao *et al.*, *Sol. Energy Mater. Sol. Cells* 141, 132 (2015). Copyright 2015 Elsevier.

images of the MWSS and DWS mc-Si wafers after normal acid etching and MacEtch processes. As shown in Figures 1.11a, b, the oval pits in the MWSS mc-Si wafer are deeper than those of the DWS mc-Si wafer. Nanoscale pyramidal structures are produced on the acidic etched wafer surface which underwent MacEtch and subsequent alkaline treatment (Figures 1.11c, d). Figure 1.11e shows the surface reflectance spectra of wafers with different surface microstructures. The authors suggested that the reflection of a nanostructured DWS mc-Si wafer can be reduced to ~5% and an optimized average reflectance is about 15.6% by considering the balance between the optical and electrical properties of the cells. Figure 1.11f gives the I-V characteristics of the as-prepared best micron- and nano-textured cells. It can be clearly seen that the efficiency of the DWS mc-Si solar cell textured by MacEtch and alkaline treatment (18.31%) is higher than that of the DWS one (17.67%) textured by normal acidic etching.

1.4.4 Copper-MacEtch Inverted Pyramid Black Silicon for Photovoltaic Solar Cells

Silver-, gold-, palladium-, and copper-MacEtch b-Si samples with porous layers suffer from serious surface/interface recombination due to their ultralarge surface areas. In 2014, Toor *et al.* prepared b-Si by copper nanoparticles catalyzed etching and demonstrate a 17.0% efficient b-Si solar cell without additional ARC [129]. Su and coauthors demonstrate that the efficiency of DWS mc-Si solar cells produced by copper-MacEtch process is greater than 19% [164]. It is interesting that Yang and coauthors demonstrated that inverted pyramid b-Si could be fabricated by one-step maskless copper-assisted texturization of the silicon surfaces [131, 165]. This inverted pyramid b-Si has the potential to avoid severe surface/interface recombination losses thanks to its big and open structure compared to conventional MacEtch b-Si. They utilized Cu nanoparticles to catalyze anisotropic etching of silicon and fabricate inverted pyramid. Random micrometer-sized inverted pyramids could be observed on the silicon surface after one-step copper-assisted etching in HF solution [166]. In recent years, MacEtch to optimize silicon surface texture are becoming increasingly significant in the industrial PV industry. Various MacEtch have made great progress in improving the efficiency of solar cells. It is important to note that the inverted pyramid structure by MacEtch is very easy to adapt to the prevailing PERC technology. The formation of silicon surface morphology by silver catalysis has been studied extensively [167–176]. Sliver-MacEtch is regarded as the initial metal-catalyzed etching because it requires an additional step called nanostructure rebuilding (NSR) process to enlarge the hole with an alkaline solution or nitric acid after etching by silver ion deposition. However, the cost of noble metal such as silver and gold particles is high and cannot adapt to the globalization trend of increasing cost reduction of solar cells. Copper as a cheap metal can also act as a catalyst in metal catalytic etching [177, 178]. Cu has previously been successfully used in industrial solar cell fabrication as an electroplated current collector metal grid and Cu impurities are weak recombination centres in silicon. At present, copper has gradually replaced silver due to low-cost in the textured process of solar cells technology [179–187]. National Renewable Energy Laboratory and Oh reported that copper could be used as a catalyst to prepare solar cells using MacEtch method of b-Si [129]. The solution of the copper depositing step in MacEtch include ammonium fluoride (NH_4F), copper sulfate ($CuSO_4$), ascorbic acid ($C_6H_8O_6$), sodium potassium tartrate ($KNaC_4H_4O_6{\bullet}4H_2O$) and methanol. It is then etched wafer at 50°C with oxidizer and HF to form pyramid structure. Zheng *et al.* proposed a new method of MacEtch by Ag/Cu co-assisted chemical etching [181]. For the co-catalysis of copper and silver, the simultaneous use of silver nitrate and copper nitrate as the main solvent of MacEtch can overcome the fact that copper does not react with silicon at room temperature. After that, Du *et al.* reported that the inverted pyramid structure could be prepared in one-step using HF solution including copper nitrate and hydrogen peroxide. With the development of copper MacEtch process technology, the improved methods are also introduced. Gao *et al.* applied both copper and nickel to the MacEtch [182]. They found the rate of chemical process was greatly improved and the morphology of the inverted pyramid structure was successfully achieved by adding nickel ions in copper etching solution. MacEtch is also applicable to ultra-thin silicon wafers that are close to tens of microns in size and are very popular in current research [183, 184]. The surface prepared by MacEtch also has excellent performance in ultra-thin silicon wafers and organic hybrid solar cells [185]. Silicon wafer thinning is the main improvement direction of future production line. Diamond-wire-sawn single-crystalline has completely replace multi-wire-slurry-sawing wafer. For the primary mass-produced solar cells, multi-wire-slurry-sawing (MWSS) wafer was used in industry. At present, the comprehensive cost of diamond wire saw is lower than that of slurry wire saw, and it has great advantages in thickness and surface damage layer of silicon wafer. Silicon wafers at solar cell levels are also getting thinner. Excellent inverted pyramid structure can be achieved not only in mono-crystalline silicon but also in multi-crystalline silicon [186]. The texture of ultrathin silicon wafers also indicates that MacEtch is applicable to different kinds of silicon wafers. These low-cost materials and methods can be perfectly adapted to the current production line and very mature PERC process. Figure 1.12a shows the SEM image of shallow inverted pyramid–like structures. With the increase of etching time, the inverted pyramids became standard with Si (111) sidewalls, as shown in Figure 1.12b. The average reflectivity is about 8% for the silicon surface etched for 10 min in mixed $Cu(NO_3)_2$, HF and H_2O_2 solution, and decreased to as low as 5% as the etching time extends to

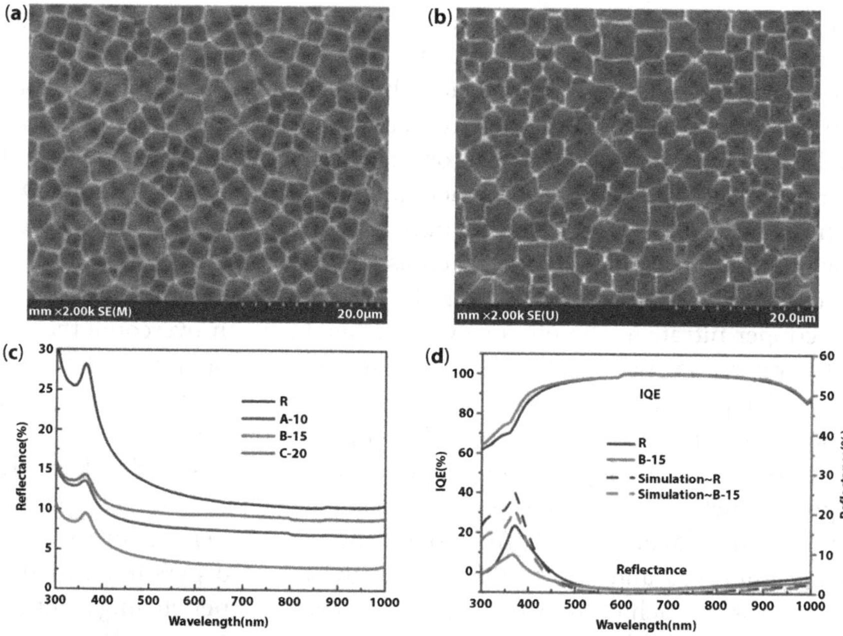

(e)

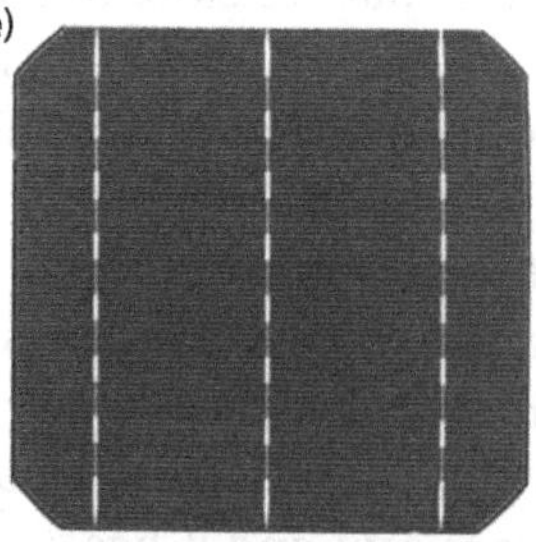

(f)

Sample	V_{oc} (mV)	J_{sc} (mA/cm²)	FF (%)	E_{ff} (%)	R_s (ohm)	
A–10	637.1	36.94	79.01	18.62	0.00219	Best
	636.3	36.95	78.67	18.52	0.00224	Average
B–15	637.7	37.47	78.84	18.87	0.00250	Best
	636.6	37.49	78.81	18.83	0.00242	Average
C–20	633.8	36.92	78.24	18.34	0.00241	Best
	634.2	36.62	78.62	18.28	0.00245	Average
R	637.0	36.88	78.22	18.40	0.00275	Best
	634.7	36.84	77.87	18.23	0.00277	Average

Figure 1.12 SEM images of the inverted pyramid structures for (a) 10 min processing and (b) 15 min processing. (c) Reflectance spectra of sample R, A-10, B-15, and C-20. (d) Reflectance spectra and IQE of sample R and B-15 with 80-nm SiNx by experiments and 3D ray tracing simulations. (e) Photograph of inverted pyramid-textured solar cell. (f) Summary of performance for the inverted pyramid structured and upright pyramid structured Si solar cells. Reprinted with permission from L. Yang *et al.*, *Sol. Energy Mater. Sol. C* 166, 121 (2017). Copyright 2017 Elsevier.

15 min (Figure 1.12c). Figure 1.12d further shows reflectance spectra with 80 nm SiN_x film by experiments and 3D ray tracing simulations. Due to the triple bounce of incident light in inverted pyramid as the authors suggested, the average reflectivity of sample B-15 is 1% lower than that of sample R over the wavelength range from 300 to 1,000 nm. The internal quantum efficiency (IQE) spectra as shown in Figure 1.12d suggested improved blue response of the inverted pyramids silicon solar cell. The authors compared the performance of Cu-etched inverted pyramid textured silicon solar cells with those of upright pyramid textured silicon cells. The 18.87% efficient copper-etched inverted pyramid b-Si solar cell with the short-circuit current density of 37.47 mA/cm^2 is reached, as shown in Figure 1.12f, slightly higher than that (18.40%) of upright pyramid textured silicon solar cell. In terms of multicrystalline silicon solar cells, Du *et al.* obtained the high-efficiency 19.49% DWS mc-Si Al-BSF solar cell by using the copper-silver co-catalytic method to obtain an inverted pyramid with excellent light trapping and passivation effects [188]. Owing to the brilliant performance of optical, b-Si has made great contribution in the field of solar energy conversion such as high-efficiency solar cells based on simple passivation trapping structure. At present, the efficiency of solar cells has exceeded 23%. As the surface structure of silicon-based solar cells, there is still a lot of room for improvement through different kinds of technologies.

1.5 Concluding Remarks

We have reviewed the origin and mechanism of MacEtch of silicon in HF solutions and its recent advances in PV application. Silicon, one of the most widely used semiconductor materials, is the foundation of high-tech society. Silicon has received a great deal of attention since it was used as a transistor to control current switches. Over the decades, the technology of various silicon materials has developed quite mature. As a special treatment for silicon wafer, b-Si can greatly improve the efficiency of photoelectric conversion of solar cells and promote the utilization of clean energy such as light energy. Despite the long history and significant achievements, the etching process remains so controversial. But nevertheless, there is widespread agreement that MacEtch of silicon originates from coupled electrochemical hole transfer mediated via noble metal particles and follows from the mixed potential theory. In addition, the explanations of copper nanoparticle catalyzed anisotropic etching of silicon are very ambiguous, this is cannot be easily understood in terms of conventional metal/silicon galvanic etching process, and the details and underlying mechanism should be systematically investigated in future. MacEtch has emerged as a powerful and low-cost surface micro/nanostructuring technique for the scalable fabrication of b-Si with excellent light management properties, which might lead to both efficiency increase and cost reduction of solar cells. SiNW-based MacEtch b-Si solar cells have aroused great interest but met daunting challenges in reducing the recombination of photo-generated charge carriers associated with the ultrahigh surface area. In contrast, MacEtch b-Si solar cells with inverted pyramidal textures, prepared by alkaline treatment modified MacEtch process or one-step copper particle catalyzed etching process, have shown its great promise for large-scale deployment due to the reduced surface area with easy controlled surface and Auger recombination. Excitingly, MacEtch DWS mc b-Si solar cells have been currently commercialized and some top solar companies are running large-scale production with success.

Different structures of b-Si have different applications. Owing to the brilliant performance of optical, b-Si has made great contribution in the field of solar energy conversion such as high-efficiency solar cells based on simple passivation trapping structure. At present, the efficiency of solar cells has exceeded 23%. As the surface structure of silicon-based solar cells, there is still a lot of room for improvement through different kinds of technologies. New sensors and light splitting water technology based on b-Si have also been increasingly applied due to the low cost and good performance of b-Si. As the most promising lithium battery to solve today's energy problems and become the core of the next generation of energy, the most likely way to greatly increase the capacity is to combine with silicon materials. However, in the process of silicon fusion into lithium batteries, there are still many technical problems that have not been fundamentally solved, such as the volume expansion of silicon and the cycle performance of batteries. Therefore, solving these problems can make the application of b-Si more extensive.

In summary, b-Si has unique excellent optical and mechanical properties which makes it suitable for many applications. However, there are still many problems to be solved in the preparation and application process of various b-Si structures. Although there are still many challenges, with great amount of in-depth research on b-Si, supplementary problems will be solved. Black silicon still has very promising applications. If the recombination issue associated with high-surface-area b-Si and environmental pollution issue associated with noble metals ions can be solved, the MacEtch technology would have the potential to completely replace traditional silicon surface texturing techniques and thereby MacEtch b-Si solar cells or energy devices of are expected to be the dominating future PV technology.

Acknowledgements

We acknowledge the contributions of our research group members and the financial support from National Natural Science Foundation of China (51972031) and Beijing Natural Science Foundation (2172030). K.Q. Peng is grateful for the Interdisciplinary support from the Beijing Normal University. K.Q. Peng supervised the project and wrote this chapter.

References

1. BP Statistical Review of World Energy 2018, BP, 2018.
2. Technology Roadmap for Photovoltaic (ITRPV): Results 2017 Including Maturity Report 2018 VDMA Photovoltaic Equipment, Munich, Germany, March 2018.
3. Battaglia, C., Cuevasb, A., Wolf, S.D., High-efficiency crystalline silicon solar cells: status and perspectives. *Energy Environ. Sci.*, 9, 1552, 2016.
4. Green, M.A., The path to 25% silicon solar cell efficiency: History of silicon cell evolution [J]. *Prog. Photovoltaics*, 17, 183, 2009.
5. Yoshikawa, K., Kawasaki, H., Yoshida, W., Silicon heterojunction solar cell with interdigitated back contacts for a photoconversion efficiency over 26%. *Nat. Energy*, 2, 5, 17032, 2017.
6. Taguchi, M., Kawamoto, K., *et al.*, HITTM Cells-High Efficiency Crystalline Si Cells with Novel Structure. *Prog. Photovoltaics*, 8, 503, 2000.
7. Taguchi, M., Kawamoto, K. *et al.*, Crystalline silicon photovoltaics: a cost analysis framework for determining technology pathways to reach baseload electricity costs. *Energy Environ. Sci.*, 5, 5874, 2012.
8. Campbell, P. and Green, M.A., Light trapping properties of pyramidally textured surfaces. *J. Appl. Phys.*, 62, 243–249, 1987.
9. Seidel, H., Csepregi, L., Heuberger, A., Baumgartel, H., Anisotropic Etching of Crystalline Silicon in Alkaline Solutions. *J. Electrochem. Soc.*, 137, 3612, 1990.
10. Adachi, D., Hernandez, J.L., Yamamoto, K., Impact of carrier recombination on fill factor for large area heterojunction crystalline silicon solar cell with 25.1% efficiency. *Appl. Phys. Lett.*, 107, 233506, 2015.
11. Blakers, A.W., Wang, A., Milne, A.M., Zhao, J., Green, M.A., 22.8% efficient silicon solar cell. *Appl. Phys. Lett.*, 55, 1363–1365, 1989.
12. Zhao, J. and Green, M.A., Optimized antireflection coatings for high-efficiency silicon solar cells. *IEEE Trans. on Electron Devices*, 38, 1925, 1991.
13. Mavrokefalos, A., Han, S.E., *et al.*, Efficient Light Trapping in Inverted Nanopyramid Thin Crystalline Silicon Membranes for Solar Cell Applications. *Nano Lett.*, 12, 6, 2792–2796, 2012.

14. Kayes, B.M., Atwater, H.A., Lewis, N.S., Comparison of the device physics principles of planar and radial p-n junction nanorod solar cells. *J. Appl. Phys.*, 97, 114302, 2005.
15. Wu, C., Crouch, C.H., Zhao, L., Carey, J.E., Younkin, R., Levinson, J.A. *et al.*, Near-unity below-band-gap absorption by microstructured silicon. *Appl. Phys. Lett.*, 78, 1850, 2001.
16. Peng, K.Q., Xu, Y., *et al.*, Aligned Single-Crystalline Si Nanowire Arrays for Photovoltaic Applications. *Small*, 1, 1062–1067, 2005.
17. Peng, K.Q., Wang, X., *et al.*, High-Performance Silicon Nanohole Solar Cells. *J. Am. Chem. Soc.*, 132, 6872–6873, 2010.
18. Liu, X., Coxon, P.R., Peters, M., Hoex, B., Cole, J.M., Fray, D.J., Black silicon: Fabrication methods, properties and solar energy applications. *Energy Environ. Sci.*, 7, 3223, 2014.
19. Otto, M., Algasinger, M., Branz, H., Gesemann, B., Gimpel, T., Füchsel, K., Käsebier, T., Koynov, S.K.S., Li, X.P., Naumann, V., Oh, J., Sprafke, A.N., Ziegler, J., Zilk, M., Wehrspohn, R.B., Black Silicon Photovoltaics. *Adv. Opt. Mater.*, 3, 147, 2015.
20. Savin, H., Repo, P., Gastrow, G.V., Orteg, P., Calle, E., Garín, M., Alcubilla, R., Black silicon solar cells with interdigitated back-contacts achieve 22.1% efficiency. *Nat. Nanotechnol.*, 10, 624, 2015.
21. Chattopadhyay, S., Huang, Y.F., Jen, Y.J., Ganguly, A., Chen, K.H., Chen, L.C., Anti-reflecting and photonic nanostructures. *Mater. Sci. Eng. R: Reports*, 69, 1, 2010.
22. Oh, J., Yuan, H.C., Branz, H.M., An 18.2%-efficient black-silicon solar cell achieved through control of carrier recombination in nanostructures. *Nat. Nanotechnol.*, 7, 743–748, 2012.
23. Dussart, R., Mellhaoui, X., Tillocher, T., Lefaucheux, P., Volatier, M., Socquet-Clerc, C., Brault, P., Ranson, P., Silicon columnar microstructures induced by an SF6/O2 plasma. *J. Phys. D: Appl. Phys.*, 38, 3395–4002, 2005.
24. Gittleman, J., Sichel, E., Lehmann, H., Widmer, R., Textured silicon: A selective absorber for solar thermal conversion. *Appl. Phys. Lett.*, 35, 742–744, 1979.
25. Huang, Y.F., Chattopadhyay, S., *et al.*, Improved broadband and quasiomnidirectional anti-reflection properties with biomimetic silicon nanostructures. *Nat. Nanotechnol.*, 2, 770–774, 2007.
26. Chow, T.P., Maciel, P.A., Fanelli, G.M., Reactive Ion Etching of Silicon in CCl_4 and HCl Plasmas. *J. Electrochem. Soc.*, 134, 1281, 1987.
27. Chen, W.H. and Hong, C.N., 0.76% absolute efficiency increase for screen-printed multicrystalline silicon solar cells with nanostructures by reactive ion etching. *Sol. Energy Mater. Sol. Cells*, 157, 48, 2016.
28. Canham, L.T., Silicon quantum wire array fabrication by electrochemical and chemical dissolution of wafers. *Appl. Phys. Lett.*, 57, 1046, 1990.
29. Striemer, C. and Fauchet, P., Dynamic etching of silicon for broadband anti-reflection applications. *Appl. Phys. Lett.*, 81, 2980, 2002.

30. Ma, L.L., Zhou, Y.C., Jiang, N., Lu, X., Shao, J., Lu, W., Hou, X.Y., Wide-band "black silicon" based on porous silicon. *Appl. Phys. Lett.*, 88, 171907, 2006.
31. Her, T.H., Finlay, R.J., Wu, C., Deliwala, S., Mazur, E., Microstructuring of silicon with femtosecond laser pulses. *Appl. Phys. Lett.*, 78, 1673, 1998.
32. Halbwax, M., Sarnet, T., Delaporte, P., Sentis, A., Etienne, H., Torregrosa, F., Vervisch, V., Perichaud, I., Martinuzzi, S., Micro and nano-structuration of silicon by femtosecond laser: application to silicon photovoltaic cells fabrication. *Thin Solid Films*, 516, 6791, 2008.
33. Nayak, B.K., Iyengar, V.V., Gupta, M.C., Efficient light trapping in silicon solar cells by ultrafast-laser-induced self-assembled micro/nano structures. *Prog. Photovolt.*, 19, 631–639, 2011.
34. Iyengar, V.V., Nayak, B.K., Gupta, M.C., Optical properties of silicon light trapping structures for photovoltaics. *Sol. Energy Mater. Sol. Cells*, 94, 2251–2257, 2010.
35. Peng, K.Q., Wu, Y., Fang, H., Zhong, X.Y., Xu, Y., Zhu, J., Uniform, Axial-Orientation Alignment of One-Dimensional Single-Crystal Silicon Nanostructure Arrays. *Angew. Chem. Int. Ed.*, 44, 2737, 2005.
36. Svetoslav, K., Brandt, M.S., Stutzmann, M., Black nonreflecting silicon surfaces for solar cells. *Appl. Phys. Lett.*, 88, 203107, 2006.
37. Morinaga, H., Suyama, M., Ohmi, T., Mechanism of Metallic Particle Growth and Metal-Induced Pitting on Si Wafer Surface in Wet Chemical Processing. *J. Electrochem. Soc.*, 141, 2834, 1994.
38. Li, X.L. and Bohn, P.W., Metal-assisted chemical etching in HF/H2O2 produces porous silicon. *Appl. Phys. Lett.*, 77, 2572, 2000.
39. Peng, K.Q., Yan, Y.J., Gao, S.P., Zhu, J., Synthesis of Large-Area Silicon Nanowire Arrays via Self-Assembling Nanoelectrochemistry. *Adv. Mater.*, 14, 1164, 2002.
40. Peng, K.Q., Hu, J.J. *et al.*, Fabrication of Single-Crystalline Silicon Nanowires by Scratching a Silicon Surface with Catalytic Metal Particles. *Adv. Funct. Mater.*, 16, 387, 2006.
41. Huang, Z.P., Fang, H., Zhu, J., Fabrication of Silicon Nanowire Arrays with Controlled Diameter, Length, and Density. *Adv. Mater.*, 19, 744, 2007.
42. Choi, W.K., Liew, T.H., Dawood, M.K., Smith, H., II, Thompson, C.V., Hong, M.H., Synthesis of Silicon Nanowires and Nanofin Arrays Using Interference Lithography and Catalytic Etching. *Nano Lett.*, 8, 3799, 2008.
43. Liu, G., Young, K.L., Liao, X., Personick, M.L., Mirkin, C.A., Anisotropic Nanoparticles as Shape-Directing Catalysts for the Chemical Etching of Silicon. *J. Am. Chem. Soc.*, 135, 12196, 2013.
44. Fang, H., Wu, Y., Zhao, J.H., Zhu, J., Silver catalysis in the fabrication of silicon nanowire arrays. *Nanotechnology*, 17, 3768–3774, 2006.
45. Chen, Y., Li, L., Zhang, C. *et al.*, Controlling kink geometry in nanowires fabricated by alternating metal-assisted chemical etching. *Nano Lett.*, 17, 1014, 2017.

46. Hochbaum, A., I, Gargas, D., Hwang, Y.J., Yang, P., Single crystalline mesoporous silicon nanowires. *Nano Lett.*, 9, 3550, 2009.
47. Chen, H., Wang, H., Zhang, X.H., Lee, C.S., Lee, S.T., Zigzag GaN/Ga2O3 heterogeneous nanowires: Synthesis, optical and gas sensing properties. *Nano Lett.*, 10, 864, 2010.
48. Weisse, J.M., Kim, D.R., Lee, C.H., Zheng, X., Vertical Transfer of Uniform Silicon Nanowire Arrays via Crack Formation. *Nano Lett.*, 11, 1300, 2011.
49. Hildreth, O.J., Lin, W., Wong, C.P., Effect of catalyst shape and etchant composition on etching direction in metal-assisted chemical etching of silicon to fabricate 3D nanostructures. *ACS Nano*, 3, 4033, 2009.
50. Hildreth, O.J., Fedorov, A.G., Wong, C.P., 3D Spirals with Controlled Chirality Fabricated Using Metal-Assisted Chemical Etching of Silicon. *ACS Nano*, 6, 10004, 2012.
51. Chang, C. and Sakdinawat, A., Ultra-high aspect ratio high-resolution nanofabrication for hard X-ray diffractive optics. *Nat. Commun.*, 5, 4243, 2014.
52. Chang, S.W., Chuang, V.P., Boles, S.T., Ross, C.A., Thompson, C.V., Ultra-high-aspect-ratio silicon nanowires fabricated using block-copolymer lithography and metal-assisted etching. *Adv. Funct. Mater.*, 19, 2495, 2009.
53. Huang, Z., Zhang, X. *et al.*, Extended Arrays of Vertically Aligned Sub-10 nm Diameter [100] Si Nanowires by Metal-Assisted Chemical Etching. *Nano Lett.*, 8, 3046, 2008.
54. Chen, C.Y., Wu, C.S., Chou, C.J., Yen, T.J., Morphological Control of Single-Crystalline Silicon Nanowire Arrays near Room Temperature. *Adv. Mater.*, 20, 3811, 2008.
55. Huang, Z., Shimizu, T. *et al.*, Ordered Arrays of Vertically Aligned [110] Silicon Nanowires by Suppressing the Crystallographically Preferred <100> Etching Directions. *Nano Lett.*, 9, 2519, 2009.
56. Geyer, N., Huang, Z., Fuhrmann, B., Grimm, S., Reiche, M., Nguyen-Duc, T., De. Boor, J., Leipner, H., Werner, P., Goesele, U., Sub-20 nm Si/Ge Superlattice Nanowires by Metal-Assisted Etching. *Nano Lett.*, 9, 3106, 2009.
57. Peng, K.Q., Zhang, M.L., Lu, A.J., Wong, N.B., Zhang, R.Q., Lee, S.T., Ordered silicon nanowire arrays via nanosphere lithography and metal-induced etching. *Appl. Phys. Lett.*, 90, 163123, 2007.
58. Garnett, E.C. and Yang, P.D., Silicon nanowire radial p-n junction solar cells. *J. Am. Chem. Soc.*, 130, 9224, 2008.
59. Tsujino, K., Matsumura, M., Nishimoto, Y., Texturization of multicrystalline silicon wafers for solar cells by chemical treatment using metallic catalyst. *Sol. Energy Mater. Sol. Cells*, 90, 100, 2006.
60. Sivakov, V., Andra, G., Gawlik, A., Berger, A., Plentz, J., Falk, F., Christiansen, S.H., Silicon Nanowire Based Solar Cells on Glass Synthesis, Optical Properties, and Cell Parameters. *Nano Lett.*, 9, 1549, 2009.
61. Liu, Y., Lai, T., Li, H., Wang, Y., Mei, Z., Lian, H., Du, X., Nanostructure Formation and Passivation of Large-Area Black Silicon for Solar Cell Applications. *Small*, 8, 1392, 2012.

62. Li, X.L., Metal Assisted Chemical Etching for High Aspect Ratio Nanostructures: A Review of Characteristics and Applications in Photovoltaics. *Current Opinion in Solid State & Mater. Sci.*, 16, 71, 2012.
63. Peng, K.Q., Wang, X., Wu, X.L., Lee, S.T., Platinum Nanoparticle Decorated Silicon Nanowires for Efficient Solar Energy Conversion. *Nano Lett.*, 9, 3704, 2009.
64. Peng, K.Q. and Lee, S.T., ChemInform Abstract: Silicon Nanowires for Photovoltaic Solar Energy Conversion. *Adv. Mater.*, 23, 198, 2011.
65. Peng, K.Q., Wang, X., Lee, S.T., Silicon nanowire array photoelectrochemical solar cells. *Appl. Phys. Lett.*, 92, 163103, 2008.
66. Peng, K.Q., Wang, X., Li, L., Hu, Y., Lee, S.T., Silicon nanowire array photoelectrochemical solar cells. *Nano Today*, 8, 75, 2013.
67. Hochbaum, A., II, Chen, R., Delgado, R.D., Liang, W., Garnett, E.C., Najarian, M., Majumdar, A., Yang, P.D., Enhanced thermoelectric performance of rough silicon nanowires. *Nature*, 451, 163, 2008.
68. Peng, K.Q., Jie, J.S., Zhang, W.J., Lee, S.T., *Appl. Phys. Lett.*, Silicon nanowires for rechargeable lithium-ion battery anodes. 93, 033105, 2008.
69. Vlad, A., Reddy, A.L.M., Ajayan, A., Singh, N., Gohy, J.F., Melinte, S., Ajayan, P.M., Roll up nanowire battery from silicon chips. *Proc. Natl. Acad. Sci. U. S. A.*, 109, 15168, 2012.
70. Peng, K.Q., Wang, X., Lee, S.T., gas sensing properties of single crystalline porous silicon nanowires. *Appl. Phys. Lett.*, 95, 243112, 2009.
71. Hwang, Y.J., Boukai, A., Yang, P.D., *Nano Lett.*, 9, 410, 2008.
72. Chang, S.W., Oh, J., Boles, S.T., Thompson, C.V., Thompson, C.V., Fabrication of silicon nanopillar-based nanocapacitor arrays. *Appl. Phys. Lett.*, 96, 153108, 2010.
73. Shen, X.J., Sun, B.Q., Liu, D., Lee, S.T., Hybrid Heterojunction Solar Cell Based on Organic–Inorganic Silicon Nanowire Array Architecture. *J. Am. Chem. Soc.*, 133, 19408, 2011.
74. Elnathan, R., Delalat, B., Brodoceanu, D., Alhmoud, H., Harding, F.J., Buehler, K., Voelcker, N.H., Maximizing Transfection Efficiency of Vertically Aligned Silicon Nanowire Arrays. *Adv. Funct. Mater.*, 25, 7215, 2015.
75. Brammer, K.S., Choi, C., Oh, S., Cobb, C.J., Connelly, L.S., Loya, M., Jin, S., Antibiofouling, sustained antibiotic release by Si nanowire templates. *Nano Lett.*, 9, 3570, 2009.
76. Chern, W., Hsu, K., Chun, I.S., de Azeredo, B.P., Ahmed, N., Kim, K.H., Zuo, J., Fang, N., Ferreira, P., Li, X., Non-lithographic Patterning and Metal-Assisted Chemical Etching for Manufacturing of Tunable Light-Emitting Silicon Nanowire Arrays. *Nano Lett.*, 10, 1582, 2010.
77. Peng, K.Q., Lu, A.J., Zhang, R.Q., Lee, S.T., Motility of Metal Nanoparticles in Silicon and Induced Anisotropic Silicon Etching. *Adv. Funct. Mater.*, 18, 3026, 2008.

78. Wang, J., Hu, Y. *et al.*, Oxidant Concentration Modulated Metal/Silicon Interface Electrical Field Mediates Metal-Assisted Chemical Etching of Silicon. *Adv. Mater. Interfaces*, 5, 1801132, 2018.
79. Liu, L., Peng, K.Q. *et al.*, Fabrication of silicon nanowire arrays by macroscopic galvanic cell-driven metal catalyzed electroless etching in aerated HF solution. *Adv. Mater.*, 26, 1410, 2014.
80. Hu, Y., Peng, K.Q., Liu, L. *et al.*, Continuous-flow Mass Production of Silicon Nanowires via Substrate-Enhanced Metal-Catalyzed Electroless Etching of Silicon with Dissolved Oxygen as an Oxidant. *Sci. Rep.*, 4, 3667, 2014.
81. Hu, Y., Peng, K.Q., Qiao, Z. *et al.*, Metal-Catalyzed Electroless Etching of Silicon in Aerated HF/H2O Vapor for Facile Fabrication of Silicon Nanostructures. *Nano Lett.*, 14, 4212, 2014.
82. Dimova-Malinovska, D., Sendova-Vassileva, M., Tzenov, N., Kamenova, M., Preparation of thin porous silicon layers by stain etching. *Thin Solid Films*, 297, 9, 1997.
83. Morinaga, H., Futatsuki, T., Ohmi, T., Fuchita, E., Oda, M., Hayashi, C., Behavior of ultrafine metallic particles on silicon wafer. *J. Electrochem. Soc.*, 142, 966, 1995.
84. Jeon, J.S., Raghavan, S., Parks, H.G., Lowell, J.K., Ali, I., Electrochemical investigation of copper contamination on silicon wafers from HF solutions. *J. Electrochem. Soc.*, 143, 2870, 1996.
85. Teerinck, I., Mertens, P.W., Schmidt, H.F., Meuris, M., Heyns, M.M., Impact of the Electrochemical Properties of Silicon Wafer Surfaces on Copper Outplating from HF Solutions. *J. Electrochem. Soc.*, 143, 3323, 1996.
86. Ohmi, T., Imaoka, T., Sugiyama, I., Kezuka, T., Metallic impurities segregation at the interface between Si wafer and liquid during wet cleaning. *J. Electrochem. Soc.*, 139, 3317, 1992.
87. Kim, J.S., Morita, H., Joo, J.D., Ohmi, T., The Role of Metal Induced Oxidation for Copper Depositionon Silicon Surface. *J. Electrochem. Soc.*, 144, 3275, 1997.
88. Ohmi, T., Imaoka, T., Kezuka, T., Takano, J., Kogure, M., Segregation and Removal of Metallic Impurity at Interface of Silicon and Fluorine Etchant. *J. Electrochem. Soc.*, 140, 811, 1993.
89. Norga, G.J., Platero, M., Black, K.A., Reddy, A.J., Michel, J., Kimerling, L.C., Mechanism of copper deposition on silicon from dilute hydrofluoric acid solution. *J. Electrochem. Soc.*, 144, 2801, 1997.
90. Reinhardt, K.A. and Kern, W., *Handbook of Silicon Wafer Cleaning Technology*, William Andrew, New York, 2008.
91. Zhang, Z.W., Lerner, M.M., Alelel, T., Keszler, D.A., Formation of a Photoluminescent Surface on n-Si by Irradiation Without an Externally Applied Potential. *J. Electrochem. Soc.*, 140, L97, 1993.
92. Ashruf, C.M.A., French, P.J., Bressers, P.M.M.C., Kelly, J.J., Galvanic porous silicon formation without external contacts. *Sens. Actuat or A-Phys.*, 74, 118, 1999.

93. Gorostiza, P., Diaz, R., Sanz, F., Morante, J.R., Different Behavior in the Deposition of Platinum from HF Solutions on n- and p-Type (100) Si Substrates. *J. Electrochem. Soc.*, 144, 4119, 1997.
94. Gorostiza, P., Díaz, R., Kulandainathan, M.A., Sanz, F., Morante, J.R., Simultaneous platinum deposition and formation of a photoluminescent porous silicon layer. *J. Electroanal. Chem.*, 469, 48–52, 1999.
95. Chattopadhyay, S., Li, X.L., Bohn, P.W., In-plane control of morphology and tunable photoluminescence in porous silicon produced by metal-assisted electroless chemical etching. *J. App. Phys.*, 91, 6134, 2002.
96. Chattopadhyay, S. and Bohn, P.W., Direct-write patterning of microstructured porous silicon arrays by focused-ion-beam Pt deposition and metal-assisted electroless etching. *J. App. Phys.*, 96, 6888, 2004.
97. Turner, D.R., On the mechanism of chemically etching germanium and silicon. *J. Electrochem. Soc.*, 107, 810, 1960.
98. Peng, K.Q. Ph.D, *Dissertation*, Tsinghua University, Beijing, China, 2004.
99. Peng, K.Q., Fang, H., Hu, J.J. *et al.*, Metal-Particle-Induced, Highly Localized Site-Specific Etching of Si and Formation of Single-Crystalline Si Nanowires in Aqueous Fluoride Solution. *Chem. Eur. J.*, 12, 7942, 2006.
100. Peng, K.Q. and Zhu, J., Simultaneous gold deposition and formation of silicon nanowire arrays. *J. Electroanal. Chem.*, 558, 35, 2003.
101. Peng, K.Q. and Zhu, J., Morphological selection of electroless metal deposits on silicon in aqueous fluoride solution. *Electrochim. Acta.*, 49, 2563, 2004.
102. Peng, K.Q., Yan, Y.J., Gao, S.P., Zhu, J., Dendrite-Assisted Growth of Silicon Nanowires in Electroless Metal Deposition. *Adv. Funct. Mater.*, 13, 127, 2003.
103. Branza, H.M., Yost, V.E., Ward, S., Kim, M., Jones, To, B., Stradins, P., Nanostructured black silicon and the optical reflectance of graded-density surfaces. *Appl. Phys. Lett.*, 94, 231121, 2009.
104. To, W.K., Tsang, C.H., Li, H.H., Huang, Z.P., Fabrication of n-type mesoporous silicon nanowires by one-step etching. *Nano Lett.*, 11, 5252, 2011.
105. Kolasinski, K.W., The mechanism of galvanic/metal-assisted etching of silicon. *Nanoscale. Res. Lett.*, 9, 432, 2014.
106. Asoh, H., Arai, F., Ono, S., Effect of noble metal catalyst species on the morphology of macroporous silicon formed by metal-assisted chemical etching. *Electrochim. Acta.*, 54, 5142, 2009.
107. Chartier, C., Bastide, S., Lévy-Clément, C., Metal-assisted chemical etching of silicon in HF–H_2O_2. *Electrochim. Acta.*, 53, 5509, 2008.
108. Geyer, N., Fuhrmann, B., Huang, Z., de Boor, J., Leipner, H.S., Werner, P., Model for the Mass Transport during Metal-Assisted Chemical Etching with Contiguous Metal Films As Catalysts. *J. Phys. Chem. C*, 116, 13446, 2012.
109. Lai, R.A., Hymel, T.M., Narasimhan, V.K., Cui, Y., Schottky Barrier Catalysis Mechanism in Metal-Assisted Chemical Etching of Silicon. *ACS Appl. Mater. Inter.*, 8, 8875, 2016.

110. Balasundaram, K., Sadhu, J.S., Shin, J.C., Azeredo, B., Chanda, D., Malik, M., Li, X., Porosity control in metal-assisted chemical etching of degenerately doped silicon nanowires. *Nanotechnology*, 23, 305304, 2012.
111. Lai, C.Q., Zheng, W., Choi, W.K., Thompson, C.V., Metal assisted anodic etching of silicon. *Nanoscale*, 7, 11123, 2015.
112. Li, L., Zhao, X.Y., Wong, C.P., Charge Transport in Uniform Metal-Assisted Chemical Etching for 3D High-Aspect-Ratio Micro- and Nanofabrication on Silicon. *ECS J. Solid State Sci. Technol.*, 4, 337, 2015.
113. Mitsugi, N. and Nagai, K., Pit Formation Induced by Copper Contamination on Silicon Surface Immersed in Dilute Hydrofluoric Acid Solution. *J. Electrochem. Soc.*, 151, G302, 2004.
114. Huang, Z., Geyer, N., Werner, P., De Boor, J., Gösele, U., Metal-assisted chemical etching of silicon: A review [J]. *Adv. Mater.*, 23, 285, 2011.
115. Lai, C.Q., Cheng, H., Choi, W.K., Thompson, C.V., Mechanics of Catalyst Motion during Metal Assisted Chemical Etching of Silicon. *J. Phys. Chem. C*, 117, 20802, 2013.
116. Yae, S., Morii, Y., Fukumuro, N., Matsuda, H., Catalytic activity of noble metals for metal-assisted chemical etching of silicon. *Nanoscale. Res. Lett.*, 7, 352, 2012.
117. Fang, H., Wu, Y., Zhao, J., Zhu, J., Silver catalysis in the fabrication of silicon nanowire arrays. *Nanotechnology*, 17, 3768, 2006.
118. Han, H., Huang, Z.P., Lee, W., Metal-assisted chemical etching of silicon and nanotechnology applications. *Nano Today*, 9, 271, 2014.
119. Smith, Z.R., Smith, R.L., Collins, S.D., Mechanism of nanowire formation in metal assisted chemical etching. *Electrochim. Acta.*, 92, 139, 2013.
120. Qu, Y., Liao, L., Li, Y., Zhang, H., Huang, Y., Duan, X., Electrically conductive and optically active porous silicon nanowires. *Nano Lett.*, 9, 4539, 2009.
121. Chiappini, C., Liu, X., Fakhoury, J.R., Ferrari, M., Biodegradable porous silicon barcode nanowires with defined geometry. *Adv. Funct. Mater.*, 20, 2231, 2010.
122. Williams, M.O., Hiller, D., Bergfeldt, T., Zacharias, M., How the Oxidation Stability of Metal Catalysts Defines the Metal-Assisted Chemical Etching of Silicon. *J. Phys. Chem. C*, 121, 9296, 2017.
123. Cao, Y., Zhou, Y., Liu, F., Zhou, Y., Zhang, Y., Liu, Y., Guo, Y., Progress and Mechanism of Cu Assisted Chemical Etching of Silicon in a Low Cu2+ Concentration Region. *ECS J. Solid State Sci. Technol.*, 4, 331, 2015.
124. Zheng, H., Han, M., Zheng, P., Zheng, L., Qin, H., Deng, L., Porous silicon templates prepared by Cu-assisted chemical etching. *Mater. Lett.*, 118, 146, 2014.
125. Toor, F., Miller, J.B., Davidson, L.M., Nichols, L., Duan, W., Jura, M.P., Black, M.R., Black, M.R. Nanostructured silicon via metal assisted catalyzed etch (MACE): chemistry fundamentals and pattern engineering. *Nanotechnology*, 27, 412003, 2016.

126. Sailor, M.J., *Porous silicon in practice: preparation, characterization and applications*, Wiley-VCH Verlag GmbH & Co. KGaA, 1-42, 2012, Online ISBN:9783527641901.
127. Zhang, X.G., Collins, S.D., Smith, R.L., Porous Silicon Formation and Electropolishing of Silicon by Anodic Polarization in HF Solution. *J. Electrochem. Soc.*, 136, 1561, 1989.
128. Ying, Z., Liao, M., Yang, X., Han, C., Li, J., Li, J., Ye, J., High-Performance Black Multicrystalline Silicon Solar Cells by a Highly Simplified Metal-Catalyzed Chemical Etching Method. *IEEE J. Photovolt.*, 6, 888, 2016.
129. Toor, F., Oh, J., Branz, H.M., Efficient nanostructured 'black' silicon solar cell by copper-catalyzed metal-assisted etching. *Prog. Photovoltaics*, 23, 1375, 2015.
130. Wang, Y., Yang, L.X., Liu, Y.P. *et al.*, Maskless inverted pyramid texturization of silicon. *Sci. Rep.*, 5, 10843, 2015.
131. Nishioka, K., Horita, S., Ohdaira, K., Matsumura, H., Antireflection sub-wavelength structure of silicon surface formed by wet process using catalysis of single nano-sized gold particle. *Sol. Energy Mater. Sol. Cells*, 92, 919, 2008.
132. Svetoslav, K., Brandt, M.S., Stutzmann, M., Stutzmann M. Black multi-crystalline silicon solar cells. *Phys. Status. Solidi (RRL)–Rapid Res. Lett.*, 1, R53, 2007.
133. Han, S.E. and Chen, G., Optical absorption enhancement in silicon nanohole arrays for photovoltaics. *Nano Lett.*, 10, 1012, 2010.
134. Wang, K.X., Yu, Z., Liu, V., Cui, Y., Fan, S., Absorption Enhancement in Ultrathin Crystalline Silicon Solar Cells with Antireflection and Light-Trapping Nanocone Gratings. *Nano Lett.*, 12, 1616, 2012.
135. Garnett, E. and Yang, P.D., Light Trapping in Silicon Nanowire Solar Cells. *Nano Lett.*, 10, 1082, 2010.
136. Srivastava, S.K., Kumar, D., Singh, P.K., Kar, M., Kumar, V., Husain, M., Excellent antireflection properties of vertical silicon nanowire arrays. *Sol. Energy Mater. Sol. Cells*, 94, 1506, 2010.
137. Boden, S.A. and Bagnall, D.M., Optimization of moth-eye antireflection schemes for silicon solar cells. *Prog. Photovolt.*, 18, 195, 2010.
138. Christesen, J.D., Zhang, X., Pinion, C.W., Celano, T.A., Flynn, C.J., Cahoon, J.F., Design Principles for Photovoltaic Devices Based on Si Nanowires with Axial or Radial p–n Junctions. *Nano Lett.*, 12, 6024, 2012.
139. Lin, C.X. and Povinelli, M.L., Optical absorption enhancement in silicon nanowire arrays with a large lattice constant for photovoltaic applications. *Opt. Express*, 17, 19371, 2009.
140. Lin, C.X. and Povinelli, M.L., *Opt. Express*, 19, A1148, 2011.
141. Lu, H. and Chen, G., Effect of aperiodicity on the b roadband reflection of silicon nanorod structures for photovoltaics. *Nano Lett.*, 7, 3249, 2007.
142. Pei, T.H., Thiyagu, S., Pei, Z., Ultra high-density silicon nanowires for extremely low reflection in visible regime. *Appl. Phys. Lett.*, 99, 153108, 2011.

143. Kelzenberg, M.D., Boettcher, S.W., Petykiewicz, J.A., Turner-Evans, D.B., Putnam, M.C., Warren, E.L., Atwater, H.A., Enhanced absorption and carrier collection in Si wire arrays for photovoltaic applications. *Nat. Mater.*, 9, 239, 2010.
144. Cao, L., Fan, P., Vasudev, A.P., White, J.S., Yu, Z., Cai, W., Brongersma, M.L., Semiconductor nanowire optical antenna solar absorbers. *Nano Lett.*, 10, 439, 2010.
145. Xie, W.Q., Oh, J., II, Shen, W.Z., Realization of effective light trapping and omnidirectional antireflection in smooth surface silicon nanowire arrays. *Nanotechnology*, 22, 065704, 2011.
146. To, W.K., Fu, J., Yang, X., Roy, V.A.L., Huang, Z., Porosification-reduced optical trapping of silicon nanostructures. *Nanoscale*, 4, 5835, 2012.
147. Spinelli, P., Verschuuren, M.A., Polman, A., Broadband omnidirectional antireflection coating based on subwavelength surface Mie resonators. *Nat. Commun.*, 3, 692, 2012.
148. Putnam, M.C., Turner-Evans, D.B., Kelzenberg, M.D., Boettcher, S.W., Lewis, N.S., Atwater, H.A., 10 μm minority-carrier diffusion lengths in Si wires synthesized by Cu-catalyzed vapor-liquid-solid growth. *Appl. Phys. Lett.*, 95, 163116, 2009.
149. Li, X., Liang, K., Tay, B.K., Teo, E.H., Morphology-tunable assembly of periodically aligned Si nanowire and radial pn junction arrays for solar cell applications. *Appl. Surf. Sci.*, 258, 6169, 2012.
150. Li, L., Peng, K.Q., Hu, B., Wang, X., Hu, Y., Wu, X.L., Lee, S.T., Broadband optical absorption enhancement in silicon nanofunnel arrays for photovoltaic applications. *Appl. Phys. Lett.*, 100, 223902, 2012.
151. Wang, X., Peng, K.Q., Pan, X.J. *et al.*, High-Performance Silicon Nanowire Array Photoelectrochemical Solar Cells through Surface Passivation and Modification. *Angew. Chem. Int. Ed.*, 50, 9861, 2011.
152. Zhang, F.Q., Peng, K.Q., Sun, R.N. *et al.*, Light trapping in randomly arranged silicon nanorocket arrays for photovoltaic applications. *Nanotechnology*, 26, 375401, 2015.
153. Sun, R.N., Peng, K.Q., Hu, B. *et al.*, Lee, S.T., Plasmon enhanced broadband optical absorption in ultrathin silicon nanobowl array for photoactive devices applications. *Appl. Phys. Lett.*, 107, 013107, 2015.
154. Fang, H., Li, X.D., Song, S., Xu, Y., Zhu, J., Fabrication of slantingly-aligned silicon nanowire arrays for solar cell applications. *Nanotechnology*, 19, 255703, 2008.
155. Zhang, C.Y., Chen, L.Z., Zhu, Y.J., Guan, Z.S., Fabrication of 20.19% Efficient Single-Crystalline Silicon Solar Cell with Inverted Pyramid Microstructure. *Nanoscale Res. Lett.*, 13, 91, 2018.
156. Jura, M.P., Miller, J.B., Forziati, J., Yim, J., Chleboski, R., Black, M., *Sol. Energy Conference and Exhibition*, 2012.
157. Huang, Z.G., Lin, X.X., Zeng, Y., Zhong, S.H., Song, X.M., Liu, C., Yuan, X., Shen, W.Z., One-step-MACE nano/microstructures for high-efficient

large-size multicrystalline Si solar cells. *Sol. Energy Mater. Sol. Cells*, 143, 302, 2015.
158. Huang, Z.G., Song, X.M., Zhong, S.H. *et al.*, 20.0% Efficiency Si Nano/Microstructures Based Solar Cells with Excellent Broadband Spectral Response. *Adv. Funct. Mater.*, 26, 1892, 2016.
159. Ye, X.Y., Zou, S., Chen, K.X. *et al.*, 18.45%-Efficient Multi-Crystalline Silicon Solar Cells with Novel Nanoscale Pseudo-Pyramid Texture. *Adv. Funct. Mater.*, 24, 6708, 2014.
160. Lin, X.X., Zeng, Y., Zhong, S.H. *et al.*, Realization of improved efficiency on nanostructured multicrystalline silicon solar cells for mass production. *Nanotechnology*, 26, 125401, 2015.
161. Xu, H.Y., Zhong, S.H., Zhuang, Y.F., Shen, W.Z., Controllable nanoscale inverted pyramids for highly efficient quasi-omnidirectional crystalline silicon solar cells. *Nanotechnology*, 29, 015403, 2018.
162. Kumar, A. and Melkote, S.N., Diamond Wire Sawing of Solar Silicon Wafers: A Sustainable Manufacturing Alternative to Lose Abrasive Slurry Sawing. *Pro. Manufact.*, 21, 549, 2018.
163. Cao, F., Chen, K.X., Zhang, J.J. *et al.*, Next-generation multi-crystalline silicon solar cells: Diamond-wire sawing, nano-texture and high efficiency. *Sol. Energy Mater. Sol. Cells*, 141, 132, 2015.
164. Zha, J.W., Wang, T., Pan, C.F. *et al.*, Constructing submicron textures on mc-Si solar cells via copper-catalyzed chemical etching. *Appl. Phys. Lett.*, 110, 093901, 2017.
165. Wang, Y., Liu, Y.P., Yang, L.X. *et al.*, Micro-structured inverted pyramid texturization of Si inspired by self-assembled Cu nanoparticles. *Nanoscale*, 9, 907, 2017.
166. Yang, L.X., Liu, Y.P., Wang, Y. *et al.*, 18.87%-efficient inverted pyramid structured silicon solar cell by one-step Cu-assisted texturization technique. *Sol. Energy Mater. Sol. C.*, 166, 121, 2017.
167. Peng, K.Q., Huang, Z.P., Zhu, J., Fabrication of large-area silicon nanowire p-n junction diode arrays. *Adv. Mater.*, 16, 73, 2004.
168. Yae, S., Kawamoto, Y., Tanaka, H., Fukumuro, N., Matsuda, H., Formation of porous silicon by metal particle enhanced chemical etching in HF solution and its application for efficient solar cells. *Electrochem. Commun.*, 5, 632, 2003.
169. Srivastava, S.K., Kumar, D., Sharma, M., Kumar, R., Singh, P.K., Silver catalyzed nano-texturing of silicon surfaces for solar cell applications. *Sol. Energy Mater. Sol. Cells*, 100, 33, 2012.
170. Yuan, H.C., Yost, V.E., Page, M.R., Stradins, P., Meier, D.L., Branz, H.M., Efficient black silicon solar cells with nanoporous anti-reflection made in a single-step liquid etch. In *Proceedings of the 34th IEEE Photovoltaic Specialists Conference (PVSC)*, Philadelphia, PA, USA, 7–12 June 2009.

171. Toor, F., Branz, H.M., Page, M.R., Jones, K.M., Yuan, H.C., multi-scale surface texture to improve blue response of nanoporous black silicon solar cells, *Appl. Phys. Lett.*, 99, 103501, 2011.
172. Jiang, Y., Shen, H., Pu, T., Zheng, C., Tang, Q., Gao, K., High efficiency multi-crystalline silicon solar cell with inverted pyramid nanostructure Jiang. *Sol. Energy*, 142, 91, 2017.
173. Lu, Y.T. and Barron, A.R., Anti-reflection layers fabricated by a one-step copper-assisted chemical etching with inverted pyramidal structures intermediate between texturing and nanopore-type black silicon. *J. Mater. Chem. A*, 2, 12043, 2014.
174. Zhong, H., Guo, A.R., Guo, G.H., Li, W., Jiang, Y.D., The Enhanced Light Absorptance and Device Application of Nanostructured Black Silicon Fabricated by Metal-assisted Chemical Etching. *Nanoscale. Res. Lett.*, 11, 322, 2016.
175. Tang, Q.T., Shen, H.L., Gao, K., Ya, H.Y., Jiang, Y., Zheng, C.F., Pu, T., Li, Y.F., Liu, Y.W., Zhang, L., Efficient light trapping of quasi-inverted nanopyramids in ultrathin c-Si through a cost-effective wet chemical method. *RSC Adv.*, 6, 96686, 2016.
176. Wu, Y., Gao, F., Wu, H.J., Liu, X.J., Zheng, X.Y., Liu, S.Z., Wang, H.S., Zhou, S.J., Li, F.X., The effects of Ag particle morphology on the antireflective properties of silicon textured using Ag-assisted chemical etching. *J. Alloy, Compd.*, 670, 156, 2016.
177. Huang, Z.P., Geyer, N., Liu, L.F., Li, M.Y., Zhong, P., Metal-assisted electrochemical etching of silicon. *Nanotechnology*, 21, 465301, 2010.
178. Lee, J.P., Choi, S., Park, S., Extremely Superhydrophobic Surfaces with Micro- and Nanostructures Fabricated by Copper Catalytic Etching. *Langmuir*, 27, 809, 2011.
179. Wang, P., Xiao, S.Q., Jia, R. *et al.*, 18.88%-efficient multi-crystalline silicon solar cells by combining Cu-catalyzed chemical etching and post-treatment process. *Sol. Energy*, 169, 153, 2018.
180. Chen, W., Liu, Y.P., Yang, L.X. *et al.*, Difference in anisotropic etching characteristics of alkaline and copper based acid solutions for single-crystalline Si. *Scientific Reports*, 8, 3408, 2018.
181. Zheng, C.F., Shen, H.L., Pu, T. *et al.*, High-Efficient Solar Cells by the Ag/Cu-Assisted Chemical Etching Process on Diamond-Wire-Sawn Multicrystalline Silicon. *IEEE J. Photovolt.*, 7, 153, 2017.
182. Gao, K., Shen, H.L., Liu, Y.W. *et al.*, Enhanced etching rate of black silicon by Cu/Ni Co-assisted chemical etching process. *Mat. Sci. Semi. Proc.*, 88, 250, 2018.
183. Tang, Q.T., Shen, H.L., Yao, H. *et al.*, Cu-assisted chemical etching of bulk c-Si: A rapid and novel method to obtain 45 μm ultrathin flexible c-Si solar cells with asymmetric front and back light trapping structures. *Sol. Energy*, 170, 263, 2018.

184. Tang, Q.T., Shen, H.L., Yao, H.Y. *et al.*, Formation mechanism of inverted pyramid from sub-micro to micro scale on c-Si surface by metal assisted chemical etching temperature. *Appl. Surf. Sci.*, 455, 283, 2018.
185. Tang, Q.T., Shen, H.L., Ya, H.Y. *et al.*, Potential of quasi-inverted pyramid with both efficient light trapping and sufficient wettability for ultrathin c-Si/PEDOT:PSS hybrid solar cells. *Sol. Energy Mater. Sol. Cells*, 169, 226, 2017.
186. Wu, J.T., Liu, Y.P., Chen, Q.S. *et al.*, The orientation and optical properties of inverted-pyramid-like structures on multi-crystalline silicon textured by Cu-assisted chemical etching. *Sol. Energy*, 171, 675, 2018.
187. Chen, J.M., Chen, C.Y., Wong, C.P., Chen, C.Y., Inherent formation of porous p-type Si nanowires using palladium-assisted chemical etching. *Applied Surface Science*, 392, 498, 2017.
188. Chen, W., Liu, Y.P., Wu, J.T. *et al.*, High-Efficient Solar Cells Textured by Cu/Ag-Cocatalyzed Chemical Etching on Diamond Wire Sawing Multicrystalline Silicon. *ACS Appl. Mater. Interfaces*, 11, 10052, 2019.

2
Alkaline Texturing

Vanesa Fano*, Juan Carlos Jimeno, José Rubén Gutiérrez and Mª Velia Rodríguez

Technological Institute of Microelectronics (TiM), University of the Basque Country (UPV/EHU), Bilbao, Spain

Abstract
Texturing is the most common technology used in the reduction of optical losses in monocrystalline silicon solar cells, in order to increase the collected photons and thus improve their efficiency. Alkaline texturing consists of the formation of square-based pyramids randomly distributed on the surface of the wafer. This chapter includes a detailed study of the texturing process, describing the factors, parameters, and issues involved. In addition, the texturing process is located in the whole manufacturing process of the solar cell, highlighting the importance of the previous steps for a suitable resultant texturing.

Keywords: Anisotropic etching, anisotropic texturing, reflectance

2.1 Introduction to Alkaline Texturing

Texturing has been widely used to reduce optical losses, being a key issue in the photovoltaic (PV) technology. The reduction of optical losses is directly related to an improvement of solar cell efficiency by increasing the photogeneration current. Anisotropic etching achieves pyramidal structures on the surface so that reflected light can be collected and light trapping improved. The oblique incidence of light on the silicon surface allows better absorption of light near the front p-n junction. In addition, the textured surface improves the internal capture of light, as light not absorbed in the first pass has further opportunities to be absorbed, being very beneficial for cells with short diffusion lengths [1–4].

**Corresponding author*: vanesa.fano@ehu.eus.

Monika Freunek Müller (ed.) Photovoltaic Manufacturing: Etching, Texturing, and Cleaning, (43–82)

The texturing of monocrystalline Si wafers with orientation (100), after aligning the surfaces in (100) crystallographic direction while ingot cutting, is performed by wet anisotropic etching, resulting in a surface with square-based pyramids randomly distributed over it, as shown in Figure 2.1, a textured surface with pyramids of different sizes randomly distributed. Texturing is usually a two-step process, where in the first step 10–20 μm of damage and contaminated material from the previous sawing step has to be removed by a concentrated alkaline solution. Subsequently, the wafers are normally textured in a solution of potassium hydroxide (KOH) or sodium hydroxide (NaOH), with Isopropyl Alcohol (IPA) at a process temperature of 80°C. Alkaline texturing is a very slow process, being about 20 minutes or longer. Once textured, the wafers are rinsed with Deionized Water (DiW) and dipped in a dilute Hydrogen Fluoride (HF) solution to remove the chemical oxide. Then, a RCA cleaning procedure is carried out; the first stage is a RCA-1 clean, also called SC-1, a $NH_4OH:H_2O_2:H_2O$ mixture for removing organic residue, and the second stage is RCA-2 or SC-2 clean, a $HCl:H_2O_2:H_2O$ mixture for removing metallic ions. Followed by theirs corresponding rinses and dipping in dilute HF, the surface becomes suitable for the next phosphorus doping process [5]. Texturing is commonly done in the early stages of the standard solar cell manufacturing process. As shown in Figure 2.2, the process begins with a saw damage removal, which prepares the surface for the subsequent texturing. This is followed by the diffusion, edge isolation and front and rear metallization steps.

Figure 2.1 Scanning electron microscopy (SEM) photograph of a textured surface.

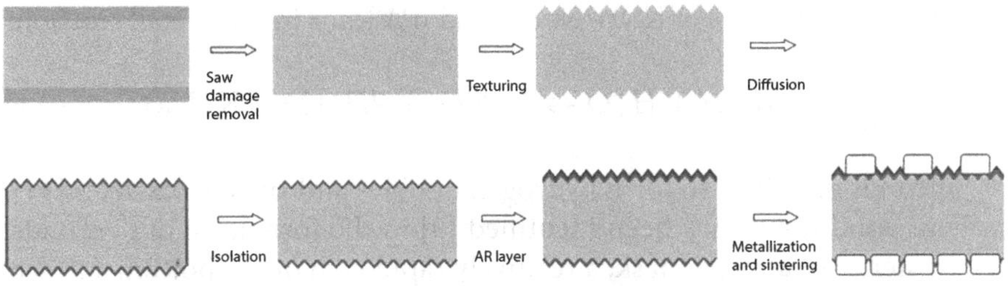

Figure 2.2 Location of the alkaline texturing within the entire process of a typical solar cell manufacturing.

Alkaline texturing is a wet etching process involving a sequence of three stages. As described in Figure 2.3, in a first phase, the reagents diffuse to the surface, where the chemical reaction takes place (second phase), in a third phase, the reaction products are desorpted from the surface to the bulk of the mixture. Any of these stages can be the limiting one, taking the control of the process [6].

In the event that wet etching is limited by the kinetics of the surface reaction, there is a tendency to promote crystallographic orientation, since surface activity depends strongly on crystallographic orientation. On the other hand, wet etching can be limited by the mass transport through the stagnant layer on the surface, it is said that is not susceptible to the nature of the surface, leading to isotropic etching [7].

Alkaline hydroxides are commonly used as reagents for silicon wet etching with a dependence on crystalline orientation [8]. The (100) planes are etched relatively faster than other planes, such as (110) and (111). This results in the intersection of the planes (111), a surface quite resistant to etching, so they are faces of the pyramids formed of a size about 3–5 μm on the exposed surface [9, 10].

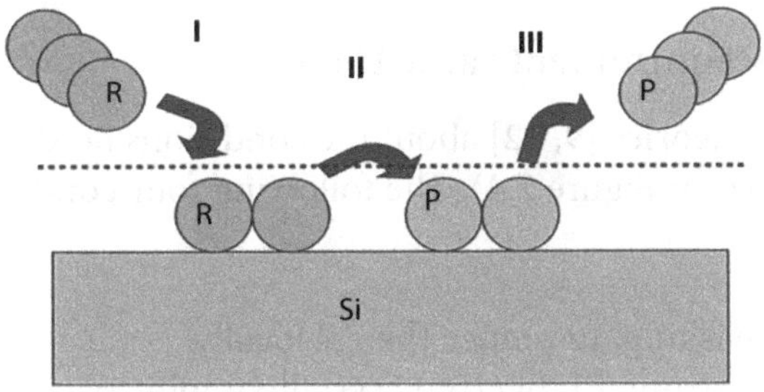

Figure 2.3 Sequence of steps involved in the wet etching [6].

A texturing mixture of KOH/NaOH and IPA leads in the overall reaction:

$$Si + 2MOH + H_2O \rightarrow M_2SiO_3 + 2H_2 \text{ (M = K, Na)} \quad (2.1)$$

Silicon is dissolved to form hydrogen bubbles and silicate particles [11].

Four conditions have been identified [9] in the formation of pyramids: the existence of a micromask, a relatively rapid etch of the bottom surface, lower etch rate (ER) of the vertices than the bottom surface, and a very low ER of the four faces. A micromask stabilizes the top of the pyramids. Several elements such as hydrogen bubbles (H_2) produced during the reaction, some of the reaction products, or some impurities present in the solution have been proposed [10] to act with micromasking effect or sometimes responsible for affecting the proportions of attack etched at specific sites: In addition to these micromasking elements, the ER of the bottom surface (100) must be relatively higher than the vertices (110) and the faces of pyramids (111).

As it has been explained, the ER of the different planes plays a fundamental role in the texturing, so it is interesting to know that the ER can be altered by rapid agitation to improve the elimination of reaction products, silicate particles, and gas bubbles or by increasing the temperature of the reactive solution. Typically, the etching rate doubles as the temperature is increased by 10°C. In addition to these factors, the choice of chemicals and their concentrations is also important [6].

Alkaline texturing as a wet etching process has certain considerations, such as the large amount of involved chemicals, so that both costs, treatment and risk aspects have to be considered. So, a great effort has been made to develop new recipes to minimize the amount of chemicals or times, always following Environment, Health, and Safety (EHS) policies.

2.2 Pyramid Formation

2.2.1 Main Requirements and Theories

There are many theories [9, 12] about the conditions hold for pyramid formation (as shown in Figure 2.4). The following four conditions have to be satisfied:

1. Micromasking, to reduce the ER locally.
2. Relative strong ER of (100) crystallographic plane, related to bottom surface.

Figure 2.4 SEM photograph of the stabilization at top and vertices of a pyramid for its formation.

3. The ER of the planes (110), which are the borders of the pyramids, must be greater than the ER of the plane (100).
4. The ER of the planes (111), faces of the pyramids, must be extremely slow.

The first condition refers to some elements acting as micromask, while the second, third, and fourth points refer to the ER of the crystallographic planes. In such a way that it is essential to explain why some planes are more likely to being etched than others. This different interaction of the dissolution with the crystallographic planes can be correlated with the bond configuration of each plane [10].

As previously stated in the first point of the conditions, there must be certain elements acting as micromasking. For years, several proposals have been made to identify these micromasking. They are proposed from non-soluble products (hydrated oxides) of the reaction of the mixture with the silicon surface to metallic ions or particles of IPA or other species able to stop locally or temporally the ER [10]. Hydrogen bubbles, which are also products of the reaction, are related to the rough of the silicon. As well as, the impurities of chemicals and the lack of dissolution of hydrated oxides are related [13]. These hydrated oxides are a form of silicate particles.

The products of the reaction, the hydrogen bubbles and the silicate particles, are declared as initiators of the pyramids [14]. However, Haiss W.

et al. [15] have established the hydrogen bubbles as the main responsible for the formation of the pyramids, provided that these hydrogen molecules remain on the surface long enough to act as a mask in the silicon etching. Each atom of dissolved silicon produces two hydrogen molecules.

According to the hypothesis of A.J. Nijdam *et al.* [16] and E. Van Veenendaal *et al.* [17], which have refer to semi-permeable particles, probably reaction products adhered to the surface as responsible for the formation of pyramids. Although these particles are attached to the surface, they allow atoms directly underneath to be etched. There is only one retard of the etching compared to atoms in the immediate vicinity. Silicate particles are created continuously and deposited at random. Some ones may be larger than others and therefore have a greater ability to slow down the etching velocity. The formed hillock is stable due to the silicate particles at its top, as well as its edges. Monte Carlo simulation [16, 17] of pyramid formation by anisotropic etching of Si(100) has confirmed that pyramid formation on the silicon surface can be explained by the presence of small permeable particles covering the top of the pyramids.

Once certain elements are established acting as a semimasking and as mentioned earlier, the second, third, and fourth points are related to the ERs of the main crystallographic planes. The susceptibility of the different crystallographic planes to the formation of the pyramids depends on their bond configurations. Surfaces (100) and (110) seem to be the most reactive. However, the plane atoms (111) are quite resistant to etching [10]. The surfaces (100) and (110) dissolve much faster than the plane (111), which constitute the sides of the pyramids [9].

In the plane (100), each atom has a dihydride termination (Figure 2.5a) and connects with two atoms of the subsurface. In the case of (111) (Figure 2.5c), the surface has monohydride termination, with three bonds to the crystal structure. These differences have a great impact on its reactivity [18].

The Si etching consists of two reaction steps. The first etching step, represented in the Figure 2.6, involves an OH^- nucleophilic attack on the Si-H bond of the surface, creating the transition state of the reaction, a pentavalent Si atom. It reacts with the H_2O to give hydrogen gas (H_2) and the surface atom is hydroxilated. The presence of the OH^- group polarizes the Si-Si backbond, as illustrated in Figure 2.7. The bond is attacked by water; the OH- is added to the positively polarized surface atom and the hydrogen to the subsurface atom. In this way, the hydrogen termination of the surface is restored.

The dihydride termination is chemically less stable than the monohydride. Thus, the reaction (1), in Figure 2.6, will be more important in

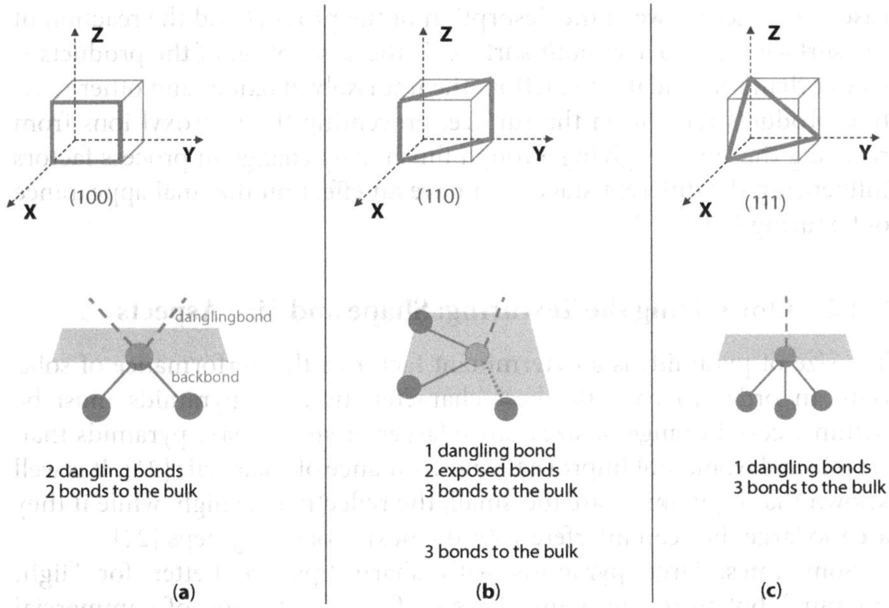

Figure 2.5 Bond configuration of (a) (100), (b) (110), and (c) (111) planes [19].

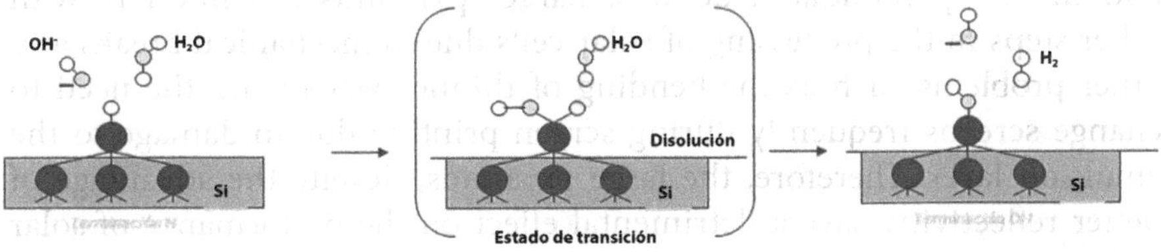

Figure 2.6 Reaction mechanism involved in the first etching step of a (100) silicon.

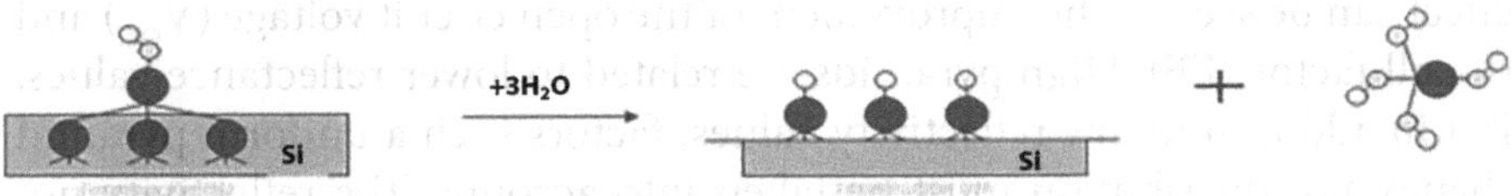

Figure 2.7 Reaction mechanism involved in the second etching step of (100) silicon [20].

the case of the surface (100), resulting in easier decoupling an atom from a plane (100). The reaction ratio (2), in Figure 2.7, should also be more important on the surface (100) [18].

Also, the texturing is determined by the relationship between the different stages, some may run faster than others or be in balance. The case of balance between the desorption of the product and the reaction of the surface leads to a smooth surface. If the desorption of the products is slower than the oxidation reaction, the undissolved oxides and other reaction products remain on the surface, preventing the hydroxyl ions from reaching the surface, giving a rough finish. Any change in process factors influencing the different stages can have an effect on the final appearance of texturing [10].

2.2.2 Optimizing the Texturing: Shape and Size Aspects

The size of pyramids is a determinant factor in the performance of solar cells. In order to have the best characteristics, the pyramids must be within a certain range of sizes, since larger or smaller size pyramids than established would not improve the performance of solar cells [21]. It is well known that if pyramids are too small, the reflectivity is high, while if they are too large they can interfere with the next processing steps [22].

Sometimes, large pyramids with sharp tips are better for "light trapping" but from the point of view of the production of commercial screen-printed solar cells [23], smaller pyramids are preferred (4–6 μm), since the metal coverage on the fingers of solar cells is better in less sharp and smaller pyramidal structures. Larger pyramids can interfere with other steps in the processing of solar cells due to mechanical breaks and other problems such as the bending of thinner wafers and the need to change screens frequently during screen printing due to damage to the emulsion layer. Therefore, the large pyramids, despite the advantage of better reflectivity, have a detrimental effect on the performance of solar cell manufacturing [24].

Small pyramids have several benefits in the processing of solar cells, enabling the diffusion process and the formation of electrical contacts. This effect can be seen in the improvement of the open circuit voltage (V_{OC}) and the fill factor (FF). High pyramids are related to lower reflectance values. But in addition to low reflectivity values, factors such a uniform pyramid shapes and distribution must be taken into account. The reflectivity not only depends on the size of the pyramids, because the area and angle of the big pyramids are close to those of small size pyramids [25].

The performance of the texturing process would be studied by means of the total hemispheric reflection of the textured surface, which is measured in a spectrometer equipped with an integrating sphere in the wavelength range of 300–1,100 nm [26]. The quality of the texturing can be easily analysed by optical micrographs under an optical microscope or by scanning electron microscopy (SEM).

Reflectivity is more influenced by the distribution of pyramids than by the absolute size of the pyramid [27]. The degradation of reflectivity for small sizes is then related to diffraction effects that become noticeable [22].

The shape and size of the pyramids formed on the silicon surface can be controlled by chemical reaction parameters such as temperature, reaction time, concentration of reagents, and the presence of additives, especially, oxidants. The ER has to be controlled in order to obtain a homogeneous distribution and size of the pyramids [28]. When the ER is high, large pyramids are formed, which can result in large surfaces that have not been textured. Furthermore, the pyramids have softened, losing their traditional tetragonal forms with defined triangular faces [16].

High temperatures improve the homogeneity of the pyramid size, because the ER of the crystallographic planes (100) and (110) increases more than (111) plane, which, reminding previous sections, satisfies several of the necessary conditions for the formation of pyramids. A higher temperature leads to the formation of larger pyramids [29].

The process time is another parameter that influences the performance of the texturing, establishing a close relationship between the average size of the pyramid and the topography of the surface. Reflectivity is reduced by increasing texturing times, but for times longer than 20 minutes, this reflection does not reduce. In process times greater than 30 minutes, some pyramids may grow at the expense of others, resulting in a non-uniform distribution, in which small and high pyramids are mixed [21]. For long etching times, the texture deteriorates considerably, so that the hemispheric reflectivity increases considerably [30]. This relationship between process time and resulted texturing quality will be explained further in Section 2.4.6.

2.3 Chemical Mixtures Used in the Alkaline Texturing

2.3.1 Overview of Chemistries Used for Alkaline Texturing

The anisotropic texturing of monocrystalline silicon is a well-known technique, based on wet etching by alkaline hydroxides, such as NaOH or KOH, and IPA. The concept of using alkaline etchants started in the microelectronics industry, where etch pits or small pyramidal hillocks showed reflection properties by the use of anisotropic etchants. Since the first mixtures consisted of alkali hydroxide and alcohol, several changes in the components of the mixture have been investigated, but keeping in mind the good performance of the initial mixture [31].

2.3.2 Most Common Chemistries

The most common mixture consists of alkali hydroxides, NaOH or KOH, IPA and sodium silicate. These solutions rely on the different ER for the planes (100) and (111), resulting in upright pyramids randomly distributed in the (100) oriented surface. Nowadays, after several work to develop the recipe, the process time has been decreased from 40 to 15–20 minutes at 80°C [32–34].

Acid etchants can also be used in texturing, but with the disadvantage of not making the silicon surface as wettable as alkaline ones, which replace hydrophobic hydrogen with hydrophilic hydrogen [35].

Dissolutions with an organic component (TMAH) can also be used. Two mechanisms compete in this type of solution: the action of OH^- ions and the blocking effect of organic molecules per ions. These mechanisms also depend on the characteristics (density of links, interatomic distances,...) of the surfaces that suffer from etching [36].

Metal hydroxides of the group I (alkali earth metals) LiOH, NaOH, KOH, present similitudes in the ER. Dissolutions of elements of the same group but with a greater atomic radius show behavior of different ER, the greater atomic radius show a greater blocking action, behaving in the same way (decrease of the ER) that molecules or organic ions. In fact, Seidel [37] proposed a mechanism and Glembocki developed it, explaining the complete etching behavior only in the case of KOH and NaOH, based on the model $R=[OH]^{x*}[H_2O]^y$. According to this model, the free water content resulting from the hydration process interferes with the etching rate curve. Ion solutions with higher atomic radius within group I and organic solutions need additional mechanisms [36].

Texturing is made in 12 wt% diluted solution and 8% IPA at 80 ± 3°C. The high anisotropy of this treatment causes the creation of several pyramids on the wafer surface, where the planes (111) are exposed, with a significant increase in surface and lower reflectivity of the material [38, 39].

It has been shown that it is important to add dissolved silicon to the texturing solution as a uniform texture is only obtained when the initial etching solution contains dissolved silicon [40]. If dissolved silicon is not added in the initial solution, the texturing homogeneity is only achieved after several processes, when there is a sufficient amount of dissolved silicon in the solution [41]. Sodium silicate is also known as water glass or soluble glass, a viscous substance whose production comes from the dissolution of substantial amounts of silicon in KOH/NaOH. The need of adding sodium silicate in fresh solutions is clear, as it has been discussed in Section 2.2 theories of pyramids formation, the reaction products of silicon play an important role in the formation of the pyramids. But unfortunately, it could be that the OH^- groups of the accumulated reaction products compete with OH^- in solution to etch the silicon atoms with triple bond located at the bottom of the pyramids and thus promote the vertical etch [42].

The solutions with significant amount of silicates tend to be not wettable surfaces. Therefore, the need to add a surface active agent, such as IPA, arises [43]. IPA is the main additive used in reactive mixtures for alkaline texturing. The interfacial energy Si/electrolyte must be lowered in order to obtain sufficient wettability of the silicon surface, in order to help the pyramids nucleation. This purpose is achieved by mixing surface active additives in the solution. The IPA increases the wettability of the Silicon surface [23].

Later sections are going to be dedicated to individual role of the NaOH concentration, the various existing theories about the role of IPA in the reaction mixture and also the role of silicates in the texturing process.

2.3.3 Other Mixtures

More recent research on texturing focuses on exploring environmental friendly and cost-effective alternative solutions, particularly by reducing the amount of IPA in solutions. The price of conventional solutions is quite high and the IPA price in particular. New process techniques focus on the use of different chemicals such as tetramethyl ammonium hydroxide (TMAH), sodium carbonate (Na_2CO_3), potassium carbonate (K_2CO_3), sodium phosphate (Na_3PO_4), and sodium hypochlorite (NaClO). Solutions based on TMAH exhibit the ability to be used for texturing. For Na_2CO_3 solutions, satisfactory reflection values have been provided in the texturing of small area silicon wafers, however for long areas it is necessary to add IPA in the process [14, 29, 44, 45].

Na_2CO_3 has a lower cost than other related compounds and their handling is not harmful to the health. Texturing with Na_2CO_3 can be adapted as an industrially cost-effective and ecologically viable alternative to conventional KOH/NaOH, IPA texturing [33, 46]. The ability to achieve optimum alkaline texturing within 20 minutes of processing with a solution of 20 wt% Na_2CO_3 and 4 wt% $NaHCO_3$ has been observed [28, 47]. Although these solutions are better than conventional ones in terms of cost, because they do not require IPA, carbonate type solutions have the small disadvantage that they can crystallize easily in the texturing process [44].

NaClO, a strong oxidizing reagent, can be used for texturing. Another application for the hot NaClO solution is the removal of organic contaminants due to its oxidizing properties. NaClO is cheaper than other etching related reagents. Although NaClO may exhibit some instability problems, the decomposition ratio of NaClO is not very important for working temperature, around 80°C–85°C. The low concentration of active chlorine in dilute solution also prevents the breakdown of NaClO. In addition to the above actions, if the process is carried out in a sealed container, the decomposition of NaClO can be reduced [44].

There are some differences between textured samples with this alternative mixture, NaClO, with respect to the samples textured in a conventional NaOH solution. The average size of the pyramids is much smaller (1–2 μm) than those textured in NaOH solution. Furthermore, while the shape of the textured pyramids in NaOH is a regular polyhedron, the edges of the textured pyramids in NaOCl are soft. Finally, the number of small particles between the textured pyramids in NaOCl is greater than that to obtain in NaOH solution [44].

2.3.4 Challenges to Develop Alkaline Texturing Recipes

Besides the need to diminish the content of IPA or look for alternative candidates, the recipe for alkaline texturing as a wet etching process has to be optimized, reducing chemicals, improving process times, from an EHS and economic point of view. In addition, in the PV industry, the process flow has changed from batch to in-line configuration. But first, the recipe must be optimized, excluding the IPA and reducing process time. A process with a duration of 40 minutes is not compatible with in-line configurations; this means that the length of the equipment is too long to be feasible.

There are other concepts concerning the process that must be taken into account when transferring the high efficiency technology obtained in the laboratory, such as the process time mentioned above, having a wide process window, the complexity of the process taken to the lower level, and the cost of the process [48].

2.4 Mechanisms of Alkaline Texturing, Important Parameters Involved in Alkaline Texturing

2.4.1 Influence of Alkaline Concentration

Texturing is closely linked to the alkaline concentration used, either KOH or NaOH, owing to the direct relations of the alkali concentration to the degree of anisotropy achieved. The dependence of the ER on crystallographic orientation changes with alkali concentration. Higher concentration leads to lower anisotropic ER [49, 50]. As the concentration increases, the ER of the plane (100) decreases slightly to half of other planes. However, at lower concentrations, the ER of (100) plane is greater than that of the plane (110). It is observed that in KOH and IPA solutions with very high concentrations of KOH, the ER proportions at high-indexed planes are comparable with V_{100} y $V_{hkl} \approx V_{h11}$ [36, 51, 52].

The abrupt drop in the ER of the (100) plane at high alkaline concentrations is a consequence of the process of hydration of the ions and the consequent decrease in the content of free water in dissolution [53].

Based on the model presented by Seidel in which the ER in silicon (R) depends on the hydroxyl ions and the concentration of free water, according to the following expression: $R = [H2O]^a \times [OH-]^b$. The ionization by alkaline hydroxide is obtained through hydration, so water is consumed. At high alkaline concentrations, the free water content decreases considerably. Given the above expression, the sharp drop in the ER is justified. As the amount of water in the solution decreases, the reaction is limited [10, 54].

On the other hand, the appearance of pyramids at low alkaline concentrations is justified by a mismatch between the stages in the silicon wet etching process, as described in Section 2.2.1 (pyramid formation). In particular, when the adsorption ratio exceeds the oxidation ratio, it results to the surface roughness. As the alkali concentration increases, the stages are balanced and the surfaces have a smooth appearance [10].

There is a relation between the alkaline concentration and the density and size of the pyramids. As several authors have highlighted [9, 10, 16, 17, 36, 49, 55, 56], lower alkali concentrations lead to small pyramids but to higher density, thereby improving surface texturing. However, when the alkali concentration increases, the amount of pyramid decreases. In the case of high concentration for KOH and IPA, the pyramid sizes are more homogeneous and relatively small. But in this case, the average reflectivity is lower than in the case of lower KOH concentrations [38]. In addition to the decrease in number of pyramidal structures in higher concentrations, these pyramids have a smaller size [57].

Silicates, reaction products, dissolve more easily in alkaline solutions, and as previously mentioned, the presence of silicate particles on the surface can promote the formation of pyramids. Thus, the surface roughness increases at low alkaline concentrations due to the increase in silicate particles. Therefore, a higher density of protuberances is expected at lower concentrations of KOH. The pyramidal protuberances are preferably formed at the edge between the hydrogen bubbles adhered to the silicon surface and the reaction mixture [16, 42]. The concentration has a significant effect on both density and in the form of pyramidal structures [42].

The evolution of the texture for different quantities of NaOH and IPA has been studied to determine the effect of the concentration on the quality of the obtained texture. The effect of alkaline concentration has been studied by comparison of the textured surface, resulting from the action of solutions in which the amount of NaOH and IPA varied, the amount of silicate remaining unchanged, until there was a zero amount of IPA in the solution. In this last part, the most surprising results were obtained, such as obtaining a texture without any IPA in the solution, for dilutions of NaOH sufficiently diluted.

After 10 minutes of texturing, it has been noted that wafers textured in solutions with a lower content of NaOH (17% NaOH) had a greater density of pyramids, as shown in Figure 2.8 [58].

Figure 2.9 represents the reflectivity values for wafers processed with dissolutions at different NaOH concentrations and times and all of them

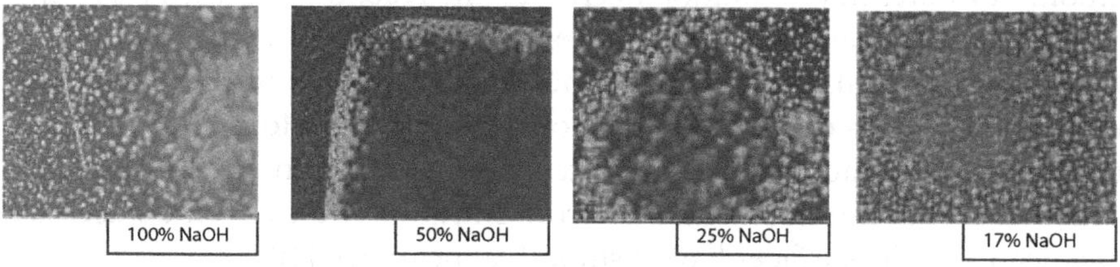

Figure 2.8 Micrographs of textured surfaces for 10 minutes in solutions containing 100%, 50%, 25%, and 17% NaOH compared to a standard textured recipe, which has been taken as reference. In addition, in all these solutions there is no IPA content [58].

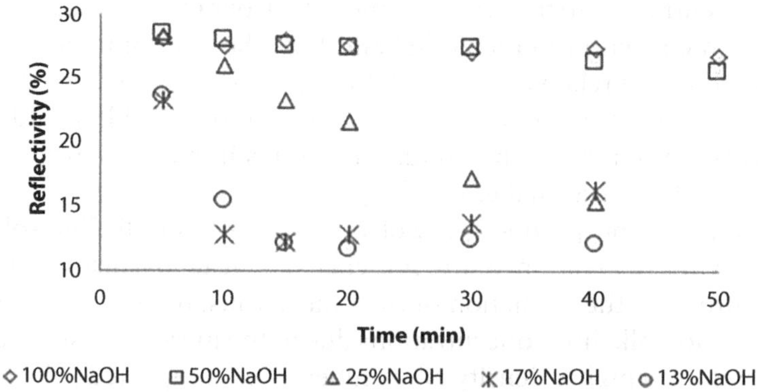

Figure 2.9 Reflectivity values at a wavelength of 725 nm, each dot means one wafer textured at time determined in the X-axis, at different percentages of NaOH, corresponding an 100% NaOH to the alkaline content of a standard recipe taken as the basis for the study. All solution were without IPA content [58].

without IPA content. It is clear that for sufficiently diluted solutions, low reflectivity values were obtained in only 10–15 minutes of processing. For diluted solutions, it is possible to have suitable reflectivity values of around 15% for contents of ≤25% NaOH. It took 40 minutes to have reflectivity of 15% when the concentration was 25% NaOH, while low reflectivity values around 12% were obtained in only 15–20 minutes for concentrations of 13%–17% NaOH. It should be highlighted again that all the dissolutions were without IPA content.

2.4.2 Additives

Some additives are added to achieve the best characteristics of the solution or even, sometimes, to overcome small inconveniences/difficulties related to the etchant employed. One of the main functions of additives is to improve the wetting characteristics of the solution used in wet etching. As shown in the following figure, the solution or drop shown on the left has good wetting characteristics, where the surface tension has been reduced, as opposed to the drop shown on the right, where the surface tension is higher and, in turn, worse wetting.

The wettability of the surface is related to the parameter S, which is detailed in the following equation as the difference between the work of adhesion (sol-liq attraction) and the work of cohesion (liq-liq attraction). In the case that S is positive, the surface is wettable, the liquid tends to extinguish on the surface of the solids, so that the contact angle between the liquid and the solids is very small, as shown in case a) of the Figure 2.10. In case b), the surface is not wettable, negative S values are obtained, because the cohesion work is greater than that of adhesion. In these poorly wettable surfaces, the contact angles between the liquid and the solid are large.

$$S = W_{adhesion} - W_{cohesion} \tag{2.2}$$

Figure 2.10 Contact angles (θ) between the liquid and the solid of two drops with different wetting characteristics, the drop on the left has good wetting ability in contrast to the one shown on the right.

Silicon is poorly wettable, due to the hydrophobic hydrogen termination of the surface. Wetting agents increase the hydrophility of the surface, providing the wetting effect without interfering with the role of the etchant. These wetting agents include surfactants and detergents, such as alcohols, ethoxylates, alkoxylates, sulfates, sulfonates, sulfoccinates, or phosphate esters and include commercial surfactant mixtures. Preferably, an alcohol such as isopropanol, methanol, or ethanol has been used [35].

2.4.2.1 IPA

IPA is the main additive used in alkaline texturing, besides acting as wetting agent or surfactant; it modulates the ER of specific crystallographic planes. There are different theories about the role of IPA in the reactant mixture. Although IPA does not participate chemically in the reaction, its presence improves, slowing down the etching. It also takes a role of oxidizing agent [59, 60]. IPA, defects or CO_3^{2-} particles have been regarded as precursors of the pyramids by some authors [61]. However, it has been observed that a purely alkaline solution can form small pyramids. With the addition of IPA or CO_3^{2-}, the formation of great pyramids is promoted. In other words, both IPA or defects and CO_3^{2-} particles should not be initiators of the pyramids but they might help form large size pyramids.

Surfaces that are not wettable or are hardly wettable present large contact angles between the solution and the substrate. For enhanced texturing, the interfacial energy between the silicon and the ionized electrolyte of the chemical solution must be reduced to achieve sufficient wettability of the silicon surface. When the silicon surface has poor wettability, an insufficient nucleation of pyramids occurs, allowing hydrogen bubbles to occupy a larger portion on the solid surface. As mentioned before, the hydrogen termination of the silicon surface makes it poorly wettable. The wetting agents used in the solution, such as IPA, increase its hydrophilicity, improve wettability, and reduce the forces that cause the bubbles to adhere to the silicon surface [35]. In this way, IPA as a surfactant improves the nucleation of pyramids. It can not only help to remove hydrogen bubbles but also to promote the formation of large pyramids [52].

The IPA addition reduces the surface tension of the solution, which influences the adhesion of bubbles to the hydrogen surface. In the absence of alcohol, the detachment of the bubbles occurs very slowly, with the corresponding development of large bubbles; therefore, large areas of the silicon surface of the reaction mixture are protected. However, the addition of IPA reduces the size of the hydrogen bubbles and these smaller bubbles can now leave the surface. The combination of changes in the size and rate of release rate of the hydrogen bubbles, together with the reactivity of adsorbed hydrogen with oxygen, leads to a considerable reduction in the hydrogen concentration on the surface [62]. IPA helps in the detachment of the bubbles, achieving a uniform texturing and a surface free of bubble spots. The hydrogenated surface atoms in the reaction go into the solution as Si(OH)4 or equivalent species, leaving the surface with a hydrogen termination. When dissolved oxygen or other oxidizing agents are present in the solution, the contact angle decreases, indicating increased adhesion between the solution and the silicon surface. Wetting properties are favored, due to an increase in hydrophilicity by the partial substitution of Si-H to give Si-OH, in the direct reaction of oxygen with Si-H [13, 63, 64].

The adsorption of IPA tends to be in specific surface bonds, in specific locations. This makes it difficult to access the reactant mixture and therefore significantly reduces the etching rate in the crystallographic planes where IPA has been adsorbed [52]. IPA mainly interacts with surface bonds of (110) plane, and does it less with surface bonds of (100) and (111) planes [55]. The compatibility of particle dimensions with the spatial configuration of the bond enhances the interaction of particles and ions, which are relatively large in size, with certain crystallographic planes. There are other factors that may influence, such as the spatial structure of the particles and their polarity (ions and polar IPA particles are adsorbed) [65].

There is a permanent competition between these IPA molecules and hydroxyl ions at specific active superficial sites. Where the IPA is adsorbed, the process of breaking bonds is temporarily inhibited. No change is observed in the plane (100) when IPA is added to the alkaline solution, so it is not likely that the IPA was adsorbed in this plane. However, it should affect the kinetic aspect of the process, as the plane (100) shows a smooth appearance after the addition of IPA in the solution. This selectivity to certain crystallographic planes is also related to the adsorption heats of water in the several planes, this heat is smaller for the surface <100> than for the <111>. The reaction requires the simultaneous adsorption of OH^- and H_2O. These, in turn, are displayed as isopropyl alcohol that reduces sites available for water adsorption, decreasing the pre-exponential factor in the expression of ER [66].

Both the texturing quality and the final cell electrical behavior are better when a high IPA content is used; the best efficiency is achieved for high concentration of KOH/NaOH and IPA. At low IPA concentration, some large pyramids between mostly small size ones are obtained. The great pyramids may have formed at the beginning of texturing and they are etched slowly. At high IPA concentration, more homogeneous pyramid size and distribution as is obtained [16]. However, if the addition of IPA is made at high alkaline concentrations, the action of IPA is limited by its low solubility at high alkaline concentrations and high pH values. Then, its adsorption on certain crystallographic planes is inhibited and consequently its function in the reaction [36].

Following the same study already shown in Section 2.4.1 [58], the evolution of pyramid formation as a function of the concentration of NaOH and IPA was reported, while the quantity of silicate incorporated remaining constant. In this case, we show the study from the point of view of the IPA concentration.

The micrographs of Table 2.1 showed that the best textured characteristics, a uniform and complete coverage of pyramids corresponded to processes placed in the diagonal (100%, 50%, and 17% IPA and NaOH content processes). The surfaces of wafers textured with these processes showed a good density of pyramids. As corroborated by the reflectivity values obtained for these wafers, 13.2%, 11.8%, and 11.1% for these processes of the diagonal.

At high IPA content, pyramid nucleation requires more time, but then, a more uniform texturing and large pyramids is observed. It has shown less difference in size of pyramid than in the case of textured wafers with less IPA.

2.4.2.2 *Other Additives*

The use of processes with IPA does not allow working at high temperatures, so the process time cannot be reduced. The IPA is a low boiling point (82°C) component, and since as the temperature of the texturing process is around 70°C–80°C, it evaporates easily and has to be redosed regularly during the process. IPA evaporation is difficult to control and the re-filling is unlikely to be optimal. The IPA concentration in the mixture should be measured and monitored continuously; their values should be closer to the initials, in order to obtain reproducible results.

A new additive should meet the requirements, providing the following advantages, which are exposed in Table 2.2 [38, 67].

There are many variables to be taken into consideration with the aim of finding an additive capable of replacing IPA. The variables include surface tension, boiling temperature, spatial conformation, and miscibility in alkaline solutions [60].

Some alternative additives to IPA have emerged, such as sodium carbonate (Na_2CO_3), potassium carbonate (K_2CO_3), and sodium phosphate (Na_3PO_4). In particular, Na_2CO_3 is a compound that provides satisfactory reflectivity values when used to texture silicon wafers with small surface area; it has lower cost than other compounds and is not considered unhealthy to handle. The Na_2CO_3 seems to be a good alternative [28].

Table 2.1 Micrographs of texturing for different concentrations (100%, 50%, and 17%) both NaOH and IPA, after 20 minutes of process. The reflectivity is shown for each wafer, which has been processed under the given concentration conditions.

	100% NaOH	**50% NaOH**	**17% NaOH**
100% IPA	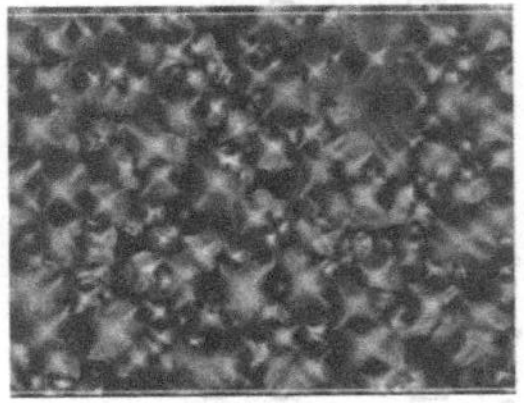 Reflectivity: 13.2%	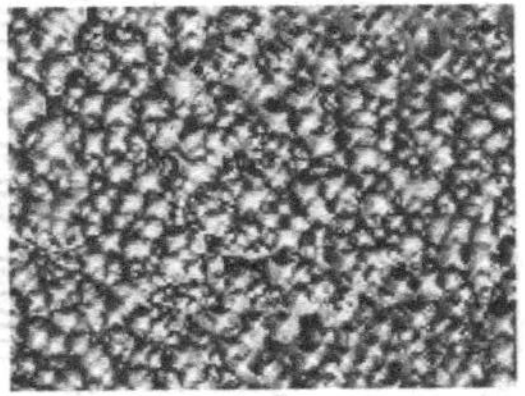 Reflectivity: 16.3%	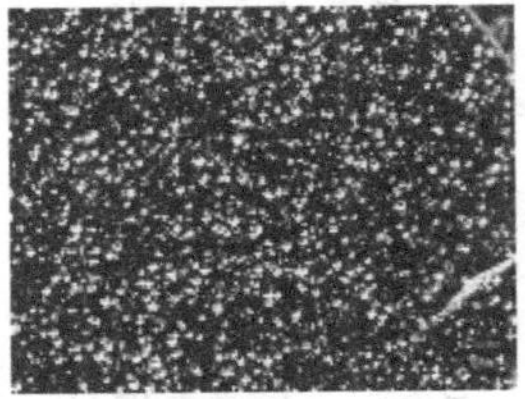 Reflectivity: 28.0%
50% IPA	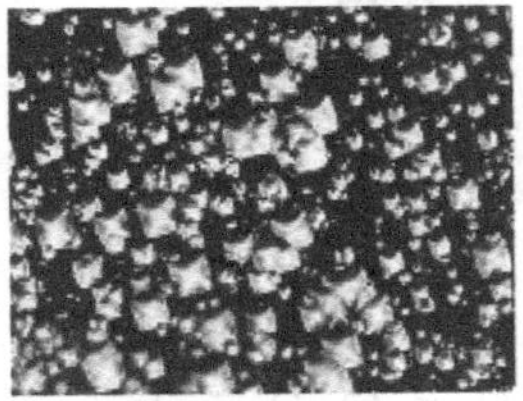 Reflectivity: 24.5%	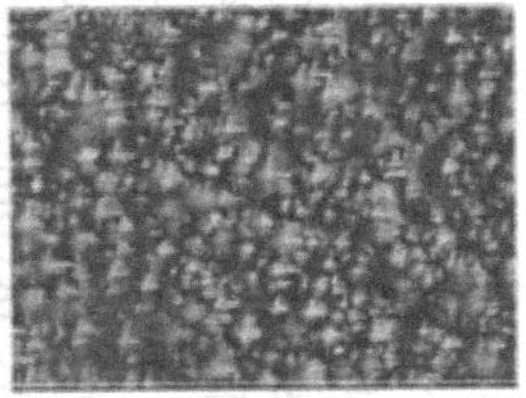 Reflectivity: 11.8%	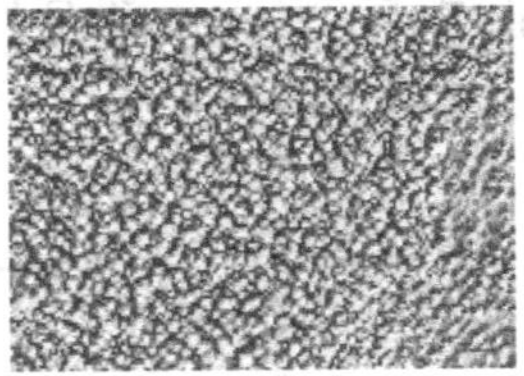 Reflectivity: 12.0%
17% IPA	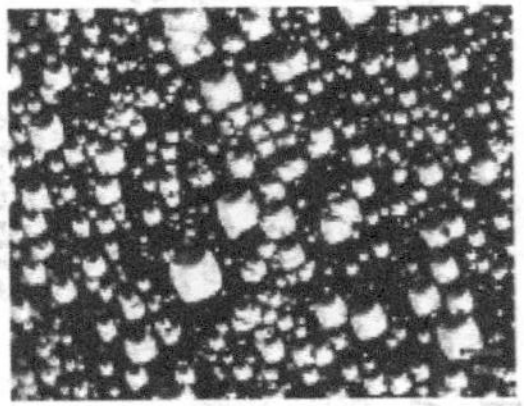 Reflectivity: 19.5%	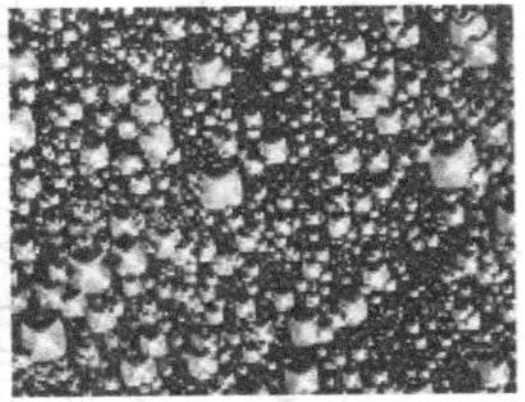 Reflectivity: 18.5%	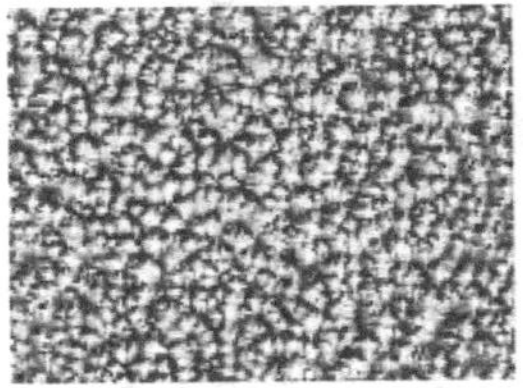 Reflectivity: 11.1%

Table 2.2 Requirements and advantages of the candidates to substitute IPA.

Requirements	**Advantages**
Higher boiling T	Less consumption and replenishment rate, more process stability and lower process time
Strong wetting function	Lower concentration
Long bath lifetime	Improved stability during the process and lower consumption and cost
Low cost and EHS complaisant	Low cost and EHS complaisant
Ability of use for in-line configurations	Suitable to the trends of PV manufacturing

Some alcohols have been proposed to replace the IPA, such as tert-butyl alcohol, TBA. Ethanol has also been used in combination with NaOCl to produce pyramidal structures [46]. Another of the substitutes tested is 1,4-cyclohexanediol in a solution with KOH and silicate, resulting in the wafer surface covered with pyramids [11, 68].

As mentioned above, surfactants normally affect surface roughness, softening the morphology, although in some cases, some orientations have softened while others have become rough. In most cases, the addition of surfactants leads to a lower contact angle and greater wettability. This facilitates the removal of bubbles and promotes the diffusion of reagents and products into the boundary layer [69]. Noncyclic diols also known as aliphatic diols have a high boiling point and solubility in alkaline solution, so they can be used as alternative additive to IPA [70].

As commercial additives, it can named ALO (Ammonyx LO), TEC (Tegotens EC 11), REWO (Rewoteric AM2 C NM), and TDO (Tegotens DO). Another commercial additives of texturing are MetaTEX 200 developed by Singulus, monoTEX F by Rena in collaboration with IMEC, and Alka-Tex by GP Solar [71].

2.4.3 Silicates

The presence of dissolved silicon is necessary to obtain a uniform texture. Experiments using a fresh solution, without dissolved silicon content, do not achieve a homogeneous texture. Whereas this purpose was achieved after two or three processes, when the amount of dissolved silicon increases, the silicate particles are proposed as responsible for the appearance of the pyramids [40, 41, 72].

They are formed all the time and deposited uniformly on the surface during etching. Some are larger than others and therefore act as a semi-mask, initiating the formation of pyramids. The particles must be permeable. So, although they remain stuck to the surface, they allow underneath atoms to be attacked, slower than the surrounding atoms around them. These particles should not completely prevent etching, and after some time, these particles should fall off [17].

E. Van Veenendaal [17] made a Monte Carlo simulation of the formation of pyramids by anisotropic etching of the plane (100) in the presence of semimasks covering only four surface atoms. Monte Carlo simulation is nothing more than a long iteration procedure to choose which surface atoms will then be extracted. The probability of an atom of remaining is determined by the number of neighboring bonds linking this atom to the crystal lattice. The atoms just below semimask are slowly etched; a small pyramid is formed with the semi-mask on its top, which expands slowly. This confirms that the pyramids are formed on the surface (100) by small permeable particles covering the top. Also, the edges of the pyramids are etched more slowly and stabilized. It is assumed that the edges of the pyramids act as sinks of small particles formed during the etching on the entire silicon surface. As a result, the edges of the pyramids are decorated with large particles. When the distance between the particles decorating the edges is not very large, the ER of the edges is considerably reduced.

Both the hydrogen bubbles and the reaction products of the silicate were proposed as elements with a masking effect, giving rise to rough surfaces. Hydrogen bubbles are formed in the reaction and remain attached to the surface, and at these points there is no reaction, resulting in pyramids. A similar assumption is made for particles, which are also a product of the reaction and their mask effect has been explained previously [14].

M.E. Dotto *et al.* [49] described the etching of NaOH on silicon, helping to understand everything related to silicate particles. Hydrolysis of the Si-H bonds followed by rapid hydrolysis of the Si-Si backbond generates an unstable HSi (OH) 3 molecule on the surface. This unstable molecule, once in solution, decomposes into $Si(OH)_4$ and H_2. In some cases, the molecules or groups of $HSi(OH)_3$ remain on the surface, blocking the process in certain areas. A summary of reactions that happen:

$$SiH_{2(surface)} + 3H_2O \rightarrow HSi\,(OH)_3 + 2H_2 \tag{2.3}$$

$$HSi(OH)_{3(surface)} \rightarrow HSi(OH)_{3(solution)} \tag{2.4}$$

$$HSi(OH)_{3(solution)} + H_2O \rightarrow Si(OH)_4 + H_2 \tag{2.5}$$

When a particle reaches a certain size, it becomes an impediment to continue the etching, forming a macroscopically visible structure that will possibly lead to zigzag. The latter reaction, in which the H terminations are re-established, is inhibited by the permanence of the reaction products on the surface. The H termination is expected to be maximum under pH conditions that maximize the dissolution rate of Si(OH)4, at a pH around 11.5 [73].

The silicate particles, as a product of the reaction, increase with time and with the processes carried out. They accumulate in the bath and adhere to the silicon surface, affecting kinetics. The increase of these silicate particles results in a decrease in etching rate. Since the silicate particles polymerize, a higher degree of polymerization implies a decrease in the etching rate [14].

As the NaOH concentration increases, the degree of polymerization is lower. When the baths are relatively fresh, only monomeric species are observed. However, as the bath ages, the degree of polymerization is higher. This effect called "solution aging" or "bath aging" is an effect caused by the increase in the number of polymer species in the solution. This phenomenon is related to the irreproducibility of etching results observed when the etchant does not refresh after a certain number of processes [17].

As explained above, the ER decreases as the bath ages. Figure 2.11 shows the behavior of the silicon ER for fresh and aged solutions and also with

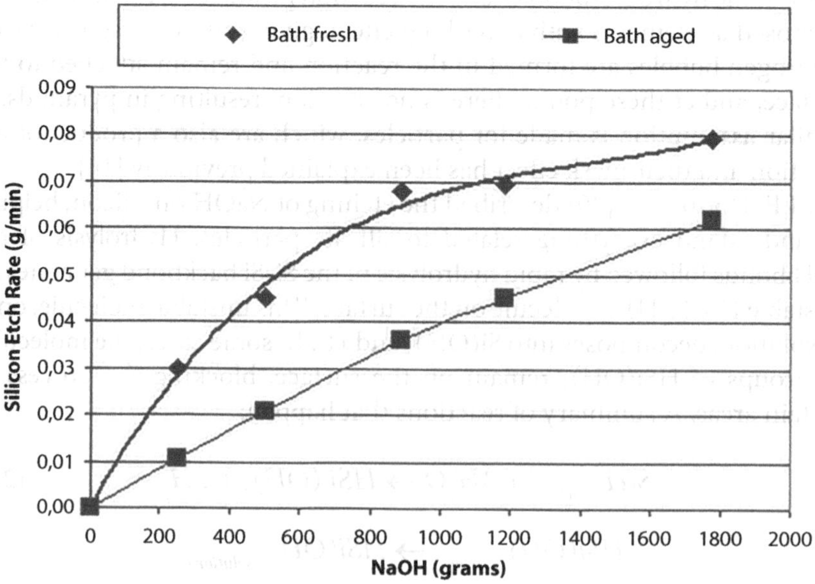

Figure 2.11 Si etch rate increasing with NaOH content for a fresh and an aged solution [74].

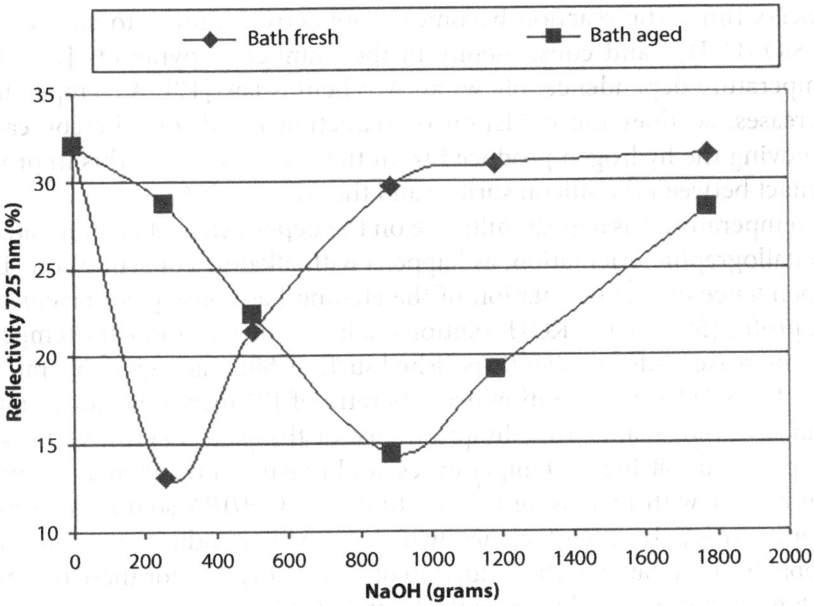

Figure 2.12 Minimum reflectivity for a fresh and aged solution [74].

different NaOH contents. Fresh baths with different NaOH contents show that the increase in NaOH implies an increase in their silicon etching rates. This trend is also visible in aged baths, which have also been carried out with different NaOH contents. However, it is observed that an aged bath shows a lower Si etching rate than a fresh bath with the same NaOH content.

Figure 2.12 shows the reflectivity values of processed wafers for fresh and aged baths in solutions with different NaOH contents, evidencing the need to increase the alkaline content when the bath is aged. As the alkaline amount increases, the degree of polymerization of the silicate particles decreases and the texturing characteristics improve. Thus, reflectivity is lower when the amount of NaOH is higher. The lower reflectivity observed for fresh solutions is around 13.11% for a NaOH content of 259 g. In other hand, for aged solutions, this lower reflectivity is moved to high NaOH value of 888 g with an also slight increasing in reflectivity value of 14.42% [74].

2.4.4 Temperature

Temperature is an important factor in the texturing. The ER gradually increases as the temperature does, facilitating the ability to lower process time. The reaction becomes more active, leading to an increase in $SiO_2(OH)_2^{2-}$ and consequently in the number of pyramids [75]. The temperature dependence follows an Arrehenius law [17]. As temperature increases, so does the oxidation or reduction of silicon, thereby easily removing the hydrogen produced from the silicon surface. This improves contact between the silicon surface and the reagents [76].

Temperature has a great influence on the dependence of etching rate on crystallographic orientation, as happens with alkaline concentration. This dependence on the orientation of the etching has consequent changes in the profile [50, 52]. For KOH solutions, at high concentration, the temperature increase leads to a slightly polished surface. Since at high temperatures the planes (110) tend to disappear, its ratio of ER increases much higher than for (100) plane. This disappearance of the planes (110), vertices of the pyramids, at higher temperatures could result in octagonal pyramids that appear with increasing temperature. In KOH/IPA solutions, at high temperatures, the ER of planes with high Miller indices also increase. Probably, it is related to the reduction of IPA adsorption for these planes at high temperatures. By lowering the temperature, the adsorption process is improved for these high Miller indices [36, 53].

As explained, the temperature has a strong influence on the morphology of the pyramids. With the temperature, the protrusions are higher and the valleys seem much smaller. Similarly, with temperature, octagonal protrusions are common [9].

2.4.5 pH

The pH of the solution is a factor with great influence on the characteristics of chemical etching, since an increase in OH^- ions on the surface involves the etching of the silicon and the dissolution of the $Si(OH)_4$ [77].

As the last reaction, the re-establishment of the H termination is inhibited by the permanence of the reaction products on the surface, and the H termination is expected to be maximum for pH conditions that maximize the $Si(OH)_4$ dissolution rate. The maximum dissolution rate of $Si(OH)_4$ is obtained at a pH of 11.5, thus the maximum H termination [73, 78].

2.4.6 Process Time

The texturing process time for a reactant mixture of NaOH with IPA is usually placed in 25–45 minutes [67]. However, according to the additive added to the mixture, the process may require more or less time, for example, in the case of KOH/CHX, the process only takes 10 minutes, since a higher temperature can be used than when IPA volatility problems arise. The process becomes three times faster than KOH/IPA [11].

A.K. Chu *et al.* [29] observed in their experiments that the coverage of pyramid on the surface was poor for times less than 15 minutes. After a process time of 15 minutes, the pyramids grew in size. The coverage of pyramid was also improved. The pyramid size grew during the processing time, indicating that the nucleation and growth of the pyramids takes place simultaneously. After 20 minutes of texturing, some pyramids grew at the expense of others, decreasing the uniformity of size, as seen in the photograph in Figure 2.13, which corresponds to a texture of 40 minutes. That means, long processing times have an improvement in the coverage of pyramids on the surface but, nevertheless, have a detrimental effect on the uniformity of the texture, as some pyramids have grown at the expense of others [67].

It has also been shown by several authors that the morphology of the pyramids can be affected for long process times. The top and vertices of large pyramids may tend to form convex edges after long process times due to the anisotropic character of itself. They can result in octagonal pyramids [44, 79].

Figure 2.14 clearly shows the lack of uniformity of the pyramids after 40 minutes of processing, some have grown at the expense of others and

Figure 2.13 SEM photograph of a Si surface with non-uniform size distribution of pyramids after 40 minutes of texturing.

Figure 2.14 SEM photograph showing the morphology of pyramids after 40 minutes of process.

in addition there has been pyramid collapse. This phenomenon consists of the destruction of the pyramid, starting from its tip to form new ones in the region that occupied the previous one.

2.4.7 Agitation

There are some kinds of agitation in the reaction solution; one is due to natural convection (density or temperature gradients) or forced convection (mechanical agitation or ultrasounds). A less common type of agitation is the result of the Marangoni effect, where two fluids with different surface tension are in contact, the gradient in surface tension results in a driving force that promotes fluid transport [49].

As far as stirring is concerned, there is a wide variety of approaches among the authors, from the authors who give importance to the presence of agitation to obtain a uniform surface finish, or a uniform texture, both in the density and in the size distribution of pyramids. Agitation would enhance the nucleation of pyramids; the growth is inhibited in favor of greater nucleation of the pyramids. Then, other authors say that agitation has an adverse effect on the texturing, obtaining smooth surfaces without pyramids, increasing the reflectivity of these surfaces.

Among the authors in favor of agitation, J.M. Kim *et al.* [32] claim that ultrasound agitation has a beneficial effect on the texturing. In order to achieve homogeneous nucleation of the pyramids, large bubbles should be avoided and for this purpose, in the case of low IPA content solutions, some agitation may be necessary.

Since the nucleation and growth of pyramids occur simultaneously. Ultrasounds seem to inhibit the growth of the pyramids in favor of nucleation; the pyramids are smaller than those formed in experiments performed without the presence of ultrasound. In addition, the uniformity in the size and shape of the pyramids improves in the presence of this type of agitation. However, when ultrasonic agitation is not applied, the growth of pyramids appears to take place even though nucleation is not provided over the entire surface of the silicon crystal.

In addition, the hydrogen bubbles produced in the reactions remain attached to the silicon surface, masking the etching by the reactants on the silicon surface. This situation inhibits homogeneous nucleation of the pyramids on the silicon surface. Then, the density of the pyramids will be lower with larger pyramids. To achieve uniform texture and smaller pyramids, it is necessary to move the hydrogen bubbles from the silicon surface. Ultrasonic agitation appears to be quite effective in moving the bubbles out of the silicon surface. Also, ultrasonic agitation promotes a faster diffusion of reaction products to the solution from the surface; a situation that increases the number of reactions and the nucleation proportion of the pyramids, enhancing the anisotropy.

J. Chen *et al.* [14] state that when there is not agitation, the reagents tend to be distributed laminarly and it appears stratified. The natural flow induced by bubbles takes on an unpredictable distribution. However, the ultrasonic agitation eliminates the stratification of the reactant mixture allowing the diffusion of the reaction mixture toward the surface.

From another point of view, there are also authors [13, 35] who establish ultrasonic agitation as an inhibitor of pyramids formation, since the adhesion of hydrogen bubbles to the surface is proposed as promoter of texturing. In the case of combining ultrasonic waves with a wetting agent in the reactive solution, there is greater uniformity in the concentration of the solution and a lower adhesion of bubbles, creating surfaces that are completely smooth and free of defects. As is known, wetting agents improve the hydrophilicity of silicon surfaces, thereby reducing the adhesion of bubbles. Ultrasonic waves help to mix the solution at the microscopic level and bubble release is improved by generating a turbulent flow in the liquid.

This decrease in the number of pyramids with ultrasonic vibrations can be explained from the perspective of the displacement of particles from the center and edges of the pyramids and perhaps completely from the substrate. The particles, after a short time of residence, rise from the center/top of the pyramid, not allowing the pyramid to grow more than a certain size [16, 17]. The power of agitation in the process affects the formation of pyramids, so that when it is strong enough, these pyramids are not formed [9].

On the subject of agitation, the authors of this chapter have an intermediate thought. Both mechanical and ultrasonic agitation are necessary, but in the case of mechanics, it should not be constant, but every 5 minutes, and in the case of ultrasound, only in the last minutes (5 minutes) with a few Hz. The reason for applying ultrasonic vibrations in these last minutes of the texturing process is to make sure that the time required for the nucleation of the pyramids and in order not to have bubble spots on the wafer surface, as shown in Figure 2.15. If ultrasound were used in the first few minutes, nucleation would not occur, as explained by Nijdam *et al.* [16], since the "semi-permeable" silicate particles would move, that is, they would not reach the silicon surface to be doing their masking function.

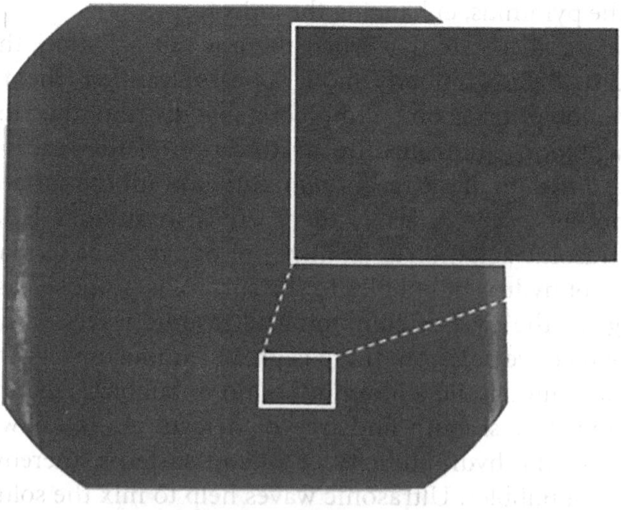

Figure 2.15 The surface of a textured wafer with bubble spots highlighted.

2.4.8 Dissolved Gases

The formation of pyramids and, in turn, the morphology of the surface are affected by the dilution of gases such as nitrogen (N_2), oxygen (O_2), and argon (Ar) in the etchant. Nitrogen and argon act in a similar way favoring the formation of pyramids; on the other hand, oxygen is related to avoiding the formation of pyramids [9]. The introduction of these diluted gases into the etchant may also have an additional stirring effect, which as explained extensively in the previous section has a beneficial effect on the quality of the resulting texturing.

In the case of alkaline solutions with IPA and saturated in oxygen, the pyramidal formation was inexistent and almost flat surfaces were obtained. Oxygen reacts rapidly with the Si-H groups generated in the solution to produce hydrogen peroxide or water. The resulting decrease in Si-H concentration at the surface reduces the reaction rate:

$$\text{Si-H} + H_2O \rightarrow \text{Si-OH} + H_2 \tag{2.6}$$

Thereby, the proportion of hydrogen bubbles is reduced. The adsorbed hydrogen decreases since it reacts with oxygen. As result, smooth surfaces are occurred. In other hand, when the alkali solutions with IPA are saturated in nitrogen, hydrogen bubbles are not inhibited. Also, the diffusion in the crystal lattice is promoted, increasing the formation of pyramids [62].

In the case of alkaline solutions with IPA saturated with oxygen, the lack of development of pyramids when adding oxygen can also be explained by the influence of the presence of an oxidizing agent on the distribution of the silicate particles and on their interaction with the silicon surface. On the other hand, saturated dissolutions in nitrogen increase the formation of pyramidal protuberances in Si(100) [17].

2.5 Surface Conditioning Prior to Alkaline Texturing

Several contaminants are present on incoming wafers and it may be reasonable knowing the origin or source of them. For example, the organic contamination is directly incorporated on the cutting step from the silicon ingot to wafers. After wire sawing process, "as-cut" wafers need to be cleaned, in order to remove remnants from the abrasives, slurries, and wire saw residues. This residue could consist of mineral oil (cutting fluids), cupper and iron oxides, and silicon carbide. The wafers from different manufacturers are contaminated with carbon and organic content on their surfaces, for example it can be presence of C=O and S-C-O groups, and triazines. These layers could impede saw damage removal and even prevent effective texturing nucleation. Therefore, these layers should be removed before texturing.

It is clear the interference of contamination in the texturing performance. Sometimes, pyramidal structures are not formed in some areas of the silicon surface, even after several consecutives cycles of saw damage etching and texturing [80–82]. Surface cleanless and damage removal are crucial for industrial production. The removal of organic residues and metallic impurity leads to an improved homogeneity in the subsequent texturing step [48, 69, 80, 83, 84].

It should be made a review of cleaning methods before the texturing step. In the manufacturing of microelectronic devices, most cleaning processes of the silicon surface have been carried out having its base in the process of standard cleaning Radio Corporation of America (RCA process), a process of wet cleaning in two stages. It is mainly based on the mixture of the active component, H_2O_2, with ammonia NH_4OH (RCAI) and hydrochloric acid HCl (RCAII). Also, a cleaning procedure based on the standard cleaning procedure RCA is used, consisting of a solution of sulfuric acid (H_2SO_4) mixed with hydrogen peroxide (H_2O_2), designated with the acronym SPM, and a solution of hydrochloric acid (HCl) mixed with a solution of H_2O_2, also referred by HPM. The wafers are rinsed in DiW between each cleaning step and immersed in Buffed Oxide Etching (BOE) to remove the native oxide layer. The proposals of RCA cleaning are risky and time consuming [69, 81, 85].

Another treatment of cleaning has been studied prior to texturing; it is a cleaning process with ethanol and IPA before treatment with HF. The surfaces of these wafers are more uniform throughout the entire area and faces of the pyramids are better defined. The homogeneity of silicon surfaces improve if the samples of the same type have been rinsed with DiW after the treatments [30].

Another pre-texturing treatment consists of cleaning silicon substrates in an ultrasonic bath with acetone, ethanol, and DiW, to remove adsorbed dust and pollution on the surface, and then, the wafers are moved to RCAI and RCAII cleaning. Finally, the native oxide layer is removed by etching in dilute HF and DiW rinse [86].

Saw damage and contamination can be effectively removed by an etching in NaOH, a concentrated NaOH solution, where around 15–20 μm of Silicon should be removed [80].

Gangopadhyay developed a pre-treatment based on NaOH/NaOCl solution at 80°C–82°C in order to remove organic and metallic contamination. In this pre-treatment, it is carried out a pre-cleaning and a saw damage cleaning simultaneously. The treatment of contaminated silicon crystalline wafers with hot NaOCl helps removing organic contaminants due to its oxidizing properties: using a hot NaOCl solution before the conventional NaOH etching. Wafers textured after a pre-treatment in NaOCl at 80°C present a uniform texturing uniform without spots on the surface. However, a conventional texturing without treatment results in a poorly textured surface [81].

For wafers without saw damage removal but with a cleaning treatment, it has been observed that the surface of these is very rough and has numerous defects. When such defects and damaged areas are present, it is difficult the proper manufacturing of the solar cell. Moreover, the efficiency of the solar cell decreases by increasing the probability of recombination of electrons and holes. For these reasons, defects and damage must be removed by NaOH etching or saw damage removal [69].

Nowadays, there are numerous suppliers of silicon wafers. Therefore, the surface conditions of as-cut wafers are different. The conditions of the wafer surface correspond directly to the different methods of cleaning used after cutting from the ingot into wafers. Wafers from different suppliers are required different texturing recipes. Currently, solar cell manufacturers face the problem of different silicon wafers suppliers and have to use many recipes of texturing. The use of different recipes results in unstable quality of texturing and lower yield [82, 87].

Saw damage removal is used as the preferred technique both to ensure that the surface conditions of the wafers prior to texturing are the same for all wafers, regardless of the wafer supplier, and to eliminate all types of contaminants and defects present in these wafers. Saw damage removal consists of a simple technique, such as etching, a concentrated solution of NaOH, which can be taken as a starting step of the texturing. This step adds time to the texturing process but assures us that the wafers will be properly textured without having to reprocess them again.

2.6 Problems Associated to Alkaline Texturing

Texturing is a fairly sensitive to the conditions and parameters in which the process is carried out. In order to avoid this extreme texturing sensitivity, it is desired that both the recipe and the entire process be as robust as possible, i.e., small variations of any parameter should not have such a significant impact on the quality of the resulting texturing. In this section, we will try to find the origin of these interferences in the quality of the resulting texture and control the corresponding parameters.

Among the problems which can occur and affect the resulting efficiency, it could be the spots of hydrogen bubbles that have remained adhered to the wafer surface for too long, where they have been formed as a reaction products, as well as areas with no pyramids or just no homogeneity of pyramidal structures that may appear.

The population distribution and pyramid size during alkaline texturing are governed by hydrogen bubbles. Hydrogen (H2) bubbles are generated during the reaction and adhere easily to the surface, causing the "pseudomasking" effect. However, some of these bubbles stuck to the surface can locally suppress the reaction that leads to the creation of pyramids. Depending on the size of the bubble, the time of residence adhered to the silicon surface is determined. In turn, it is this residence time that determines the marks of the bubbles remaining on the surface. Therefore, controlling the distribution and size of these hydrogen bubbles is the key to obtaining uniform, low-reflectivity surfaces.

In the event that large bubbles results from the reaction, it leads to a non-uniform, with several pyramid sizes. In order to avoid the formation of these large bubbles and to control the residence time of the bubbles on the surface, two solutions are proposed, such as applying forced convection by external agitation (mechanical agitation or ultrasounds) and the use of a surfactant to increase the hydrophilic nature of the Si surface. Several factors, like alkaline concentration and temperature, can influence the adhesion of the bubble on the silicon surface [40]. Concerning the influence of NaOH concentration on bubble adhesion, high concentrations lead to the formation of small bubbles and their quick release. Bubble life and average bubble size increase at lower NaOH concentrations when more silicate particles are present [15].

Sometimes, the wafers may show bubble spots with a white shade due to silicate particles deposited on the edge of the bubble. As mentioned in previous sections, when baths are aged, the particles are larger, polymerize, and precipitate more easily. On the other hand, when the baths are relatively fresh, the silicate particles are only monomeric species. Polymeric silicate species change the etching rate of silicon. As the concentration of NaOH increases, the degree of polymerization decreases and the rate of silicon etching increases [88].

In this chapter, the relevance of the use of IPA has been referred on several times, to obtain a uniform texture. Because of its problem of high volatility in temperatures around 80°C, efforts have been made to developed recipes with a low IPA content or with other additives.

Although wet chemical processing is widely employed in PV manufacturing due to its low cost, there is a trend to develop these wet chemical processes. The trial consists of reducing cost and environmental impact of these kinds of processes, through reducing chemicals and time process, simplifying processes, or even usage of non-toxic and low cost chemicals, and finally the transferring from batch to inline configurations according to trends in the current manufacturing flow.

Once the process has been optimized with all these exposed challenges achieved, texturing in short time, with low chemical consumption, and its transferring from batch technology to inline configuration, a new challenge emerges, the establishment of a "process window" providing the more stability and a repetitive recipe.

References

1. Dale, B. and Rudenberg, H., Photovoltaic conversion- high efficiency silicon solar cells. *Proceedings of the 14th Annual Power Sources Conference U.S. Army Signal Research and Development Lab*, pp. 22–25, 1960.
2. Campbell, P., Enhancement of light absorption from randomizing and geometric textures. *J. Opt. Soc. Am. (B)*, 10, 2410–2415, 1993.
3. Sopori, B.L., Reflection characteristics of textured plycrystalline silicon substrates for solar cells. *Sol. Cells*, 25, 15–26, 1988.
4. Luque, A. and Hegedus, S., *Handbook of Photovoltaic Science and Engineering*, pp. 255–303, John Wiley & Sons, West Sussex. England, 2002.
5. Barrio, R., Gonzalez, N., Carabe, J., Gandía, J.J., Optimization of NaOH texturisation process of silicon wafers for heterojunction solar-cells applications. *Sol. Energy*, 86, 845–854, 2012.
6. Ghandhi, S.K., *VLSI Fabrication principles; Silicon and Gallium Arsenide*, 2nd Edition, Wiley Interscience, New York, 1994.
7. Kelly, J.J. and Philipsen, H.G.G., Anisotropy in the wet-etching of semiconductors. *Curr. Opin. Solid State Mater. Sci.*, 9, 84–90, 2005.
8. Kovacs, G., Maluf, N.I., Petersen, K.E., Bulk micromachining of silicon. *Proc. IEEE*, 86, 8, 1536–1551 1998.
9. Suarez, M.P., Mirabella, D.A., Aldao, C.M., Dynamics of hillocks formation during wet etching. *J. Mol. Catal. A: Chem.*, 281, 230–236, 2007.
10. Zubel, I. and Kramkowska, M., The effect of isopropyl alcohol on etching rate and roughness of (100) Si surface etched in KOH and TMAH solutions. *Sens. Actuators A*, 93, 138–147, 2001.
11. Birmann, K., Zimmer, M., Rentsch, J., Fast alkaline etching of monocrystalline wafers in KOH/CHX. *23rd European Photovoltaic Solar Energy Conference*, Valencia, pp. 1608–1611, 2008.
12. Cheng, D., Gosalvez, M.A., Hori, T., Sato, K., Shikida, M., Improvement in smoothness of anisotropically etched silicon surfaces: effects of surfactant and TMAH concentrations. *Sens. Actuators A*, 125, 415–421, 2006.

13. Baum, T., Satherley, J., Schiffrin, D.J., Contact angle, gas bubble detachment and surface roughness in the anisotropic dissolution of Si (100) in aqueous KOH. *Langmuir*, 14, 10, 2925–2928, 1998.
14. Chen, J., Liu, L., Tan, Z., Jiang, Q., Fang, H., Xu, Y., Liu, Y., Study of anisotropic etching of (100) Si with ultrasonic agitation. *Sens. Actuators A*, 96, 152–156, 2002.
15. Haiss, W., Raisch, P., Bitsch, L., Nichols, R.J., Xia, X., Kelly, J.J., Schiffrin, D.J., Surface termination and hydrogen bubble adhesion on Si(100) surfaces during anisotropic dissolution in aqueous KOH. *J. Electroanal. Chem.*, 597, 1–12, 2006.
16. Nijdam, A.J., Van Veenendaal, E., Cuppen, H.M., Van Suchtelen, J., Reed, M.L., Gardeniers, J.G.E., Van Enckervort, W.J.P., Vlieg, E., Elwenspoek, M., "Formation and stabilization of pyramidal etch hillocks on silicon {100} in anisotropic etchants: Experiments and Monte Carlo simulation. *J. Appl. Phys.*, 89, 7, 4113–4123 2001.
17. Van Veenendaal, E., Sato, K., Shikida, M., Nijdam, A.J., Van Suchtelen, J., Micro-morphology of single crystalline silicon surfaces during anisotropic wet chemical etching in KOH: velocity source forests. *Sens. Actuators A*, 93, 232–242, 2001.
18. Kelly, J.J. and Philipsen, H.G.G., Anisotropy in the wet-etching of semiconductors. *Curr. Opin. Solid State Mater. Sci.*, 9, 84–90, 2005.
19. Yu, B. and Qian, L., Effect of crystal plane orientation on the friction-induced nanofabrication on monocrystalline silicon. *Nanoscale Res. Lett.*, 8, 137, 2013.
20. Gonsalvez, M.A., Sato, K., Foster, A.S., Nieminen, R.M., Tanaka, H., An atomistic introduction to anisotropic etching. *J. Micromech. Microeng.*, 17, S1–S26, 2007.
21. Ximello, N., Dastgeheib-Shirazi, A., Scholz, S., Hahn, G., Influence of pyramid size of chemically textured silicon wafers on the characteristics of industrial solar cells. *25nd EUPVSEC*, Valence, 2010.
22. Llopis, F., Tobías, I., Scholz, S., Hahn, G., Influence of texture feature size on the optical performance of silicon solar cells. *Prog. Photovolt. Res. Appl.*, 13, 27–36, 2005.
23. Gangopadhyay, U., Kim, K., Dhungel, S.K., Basu, P.K., Yi, J., Low cost texturization of large area crystalline silicon solar cells using hydrazine monohydrate for industrial use. *Renewable Energy*, 31, 1906–1915, 2006.
24. Gangopadhyay, U., Kim, K.H., Dhungel, S.K., Manna, U., Basu, P.K., Banerjee, M., Saha, H., Yi, J., A novel low cost texturization method for large area commercial mono-crystalline silicon solar cells. *Sol. Energy Mater. Sol. Cells*, 90, 20, 3557–3567, 2006.
25. Lien, S.-Y., Yang, Ch-H., Hsu, Ch-H., Lin, Y.-S., Wang, Ch-Ch., Wuu, D.-S., Optimization of textured structure on crystalline silicon wafer for heterojunction solar cell. *Mater. Chem. Phys.*, 133, 63–68, 2012.

26. Basu, P.K., Khanna, A., Hameiri, Z., The effect of front pyramid heights on the efficiency of homogeneously textured inline-diffused screen-printed monocrystalline silicon wafer solar cells. *Renewable Energy*, 78, 590–598, 2015.
27. Kegel, J., Angermann, H., Stürzebecher, U., Stegemann, B., IPA-free texturization of n-type Si wafers: correlation of optical, electronic and morphological surface properties. *Energy Procedia*, 38, 833–842, 2013.
28. Marrero, N., Gonzalez-Diaz, B., Guerrero-Lemus, R., Borchet, D., Hernandez-Rodriguez, C., Optimization of sodium carbonate texturization on large-area crystalline silicon solar cells. *Sol. Energy Mater. Sol. Cells*, 91, 1943–1947, 2007.
29. Chu, A.K., Wang, J.S., Tsai, Z.Y., Lee, C.K., A simple and cost-effective approach for fabricating pyramids on crystalline silicon wafers. *Sol. Energy Mater. Sol. Cells*, 93, 8, 1276–1280, 2009.
30. Barrio, R., Gonzalez, N., Cárabe, J., Gandía, J.J., Texturisation of Cz and Fz monocrystalline Silicon waters for a-Si/c-Si heterojuction solar cells. *25rd European Photovoltaic Solar Energy Conference*, Valencia, pp. 1621–1623, 2010.
31. Lee, D.B., Anisotropic etching of Silicon. *J. Appl. Phys.*, 40, 4569, 1969.
32. Kim, J.M. and Kim, Y.K., The enhancement of homogeneity in the textured structure of silicon crystal by using ultrasonic wave in the caustic etching process. *Sol. Energy Mater. Sol. Cells*, 81, 239–247, 2004.
33. Melnyk, I., Wefringghaus, E., McCann, M., Helfricht, A., Hauser, A., Fath, P., Na2CO3 as alternative to NaOH/IPA for texturisation of monocrystalline silicon. *19th European Photovoltaic Solar Energy Conference*, Paris, pp. 1090–1093, 2004.
34. King, D.L. and Buck, M.E., Experimental optimization of an anisotropic etching process for random texturization of silicon solar cells. *22nd IEEE Photovoltaic Specialists Conference*, vol. 303, 1991.
35. Hembree, D.R. and Akram, S., Method and apparatus for ultrasonic wet etching of silicon. *Patent. US*, 6, 224, 713 B1, 2001.
36. Zubel, I., Barycka, I., Kotowska, K., Kramskowska, M., Silicon anisotropic etching in alkaline solutions IV. The effect of organic and inorganic agents on silicon anisotropic etching process. *Sens. Actuators A*, 87, 163–171, 2001.
37. Seidel, H., Csepregi, L., Heuberger, A., Baumgärtel, H., Anisotropic etching of crystalline silicon in alkaline solutions: I. Orientation dependence and behavior of passivation layers. *J. Electrochem. Soc.*, 137, 11, 3612–3626, 1990.
38. Birmann, K., Zimmer, M., Rentsch, J., Controlling the surface tension of alkaline etching solutions. *24th European Photovoltaic Solar Energy Conference*, Hamburg, pp. 1744–1747, 2009.
39. Stefancich, M., Butturi, M., Vincenzi, D., Martinelli, G., Mechanical effects of chemical etchings on monocrystalline silicon for photovoltaic use. *Sol. Energy Mater. Sol. Cells*, 69, 371–377, 2001.

40. Papet, P., Nichiporuk, O., Kaminski, A., Rozier, Y., Kraiem, J., Lelievre, J.-F., Chaumartin, A., Fave, A., Lemiti, M., Pyramidal texturing of silicon solar cell with TMAH chemical anisotropic etching. *Sol. Energy Mater. Sol. Cells*, 90, 2319–2328, 2006.
41. Thesis of D. Iencinella, in: *Appendix A.- Silicon anisotropic etching for solar cell substrates using TMAH.*
42. Shah, I.A., Van der Wolf, B.M.A., Van Enckevort, W.J.P., Vlieg, E., Wet Chemical Etching of Silicon {111}: Autocatalysis in Pit Formation. *J. Electrochem. Soc.*, 155, 3, J79–J84, 2008.
43. Edwards, M., Bowden, S., Das, U., Burrows, M., Effect of texturing and surface preparation on lifetime and cell performance in heterojunction silicon solar cells. *Sol. Energy Mater. Sol. Cells*, 92, 1373–1377, 2008.
44. Sun, L. and Tang, J., A new texturing technique of monocrystalline silicon surface with sodium hypochlorite. *Appl. Surf. Sci.*, 255, 22, 9301–9304, 2009.
45. Iencinella, D., Centurioni, E., Rizzoli, R., Signan, F., An optimized texturing process for silicon solar cell substrates using TMAH. *Sol. Energy Mater. Sol. Cells*, 87, 725–732, 2005.
46. Marrero, N., Gonzalez Diaz, B., Borchet, D., Hernández Rodríguez, C., Guerrero Lemus, R., Study of sodium carbonate solutions for low cost texturization on large area crystalline silicon solar cells. *19th European Photovoltaic Solar Energy Conference*, Paris, France, pp. 884–886, 2004.
47. Montesdeoca-Santana, A., Jimenez-Rodríguez, E., Gonzalez-Díaz, B., Borchet, D., Guerrero-Lemus, R., Ultra-low concentration Na2CO3/NaHCO3 solution for texturization of crystalline silicon solar cells. *Prog. Photovoltaic Res. Appl.*, 20, 191–196, 2012.
48. Rentsch, J., Ackermann, R., Birmann, K., Furtwarngler, H., Haunschild, J., Kästner, G., Neubauer, R., Nievendick, J., Oltersdorf, A., Rein, S., Schütte, A., Zimmer, M., Preu, R., Wet chemical processing for c-Si solar cells-status and perspectives. *24th European Photovoltaic Solar Energy Conference*, Hamburg, pp. 1113–1117, 2009.
49. Dotto, M.E.R. and Kleinke, M.U., Scaling laws in etched Si surfaces. *Phys. Rev. B*, 65, 245353, 2002.
50. Sato, K., Shikida, M., Matsushima, Y., Yamashiro, T., Asaumi, K., Iriye, Y., Yamamoto, M., Characterization of orientation-dependent etching properties of single-crystal silicon: effects of KOH concentration. *Sens. Actuators A*, 64, 87–93, 1998.
51. Shikida, M., Sato, K., Tokoro, K., Uchikawa, D., Differences in anisotropic etching properties of KOH and TMAH solutions. *Sens. Actuators*, 80, 179–188, 2000.
52. Sundaran, K.B., Vijayakumar, A., Subramanian, G., Smooth etching of silicon using TMAH and Isopropyl alcohol for MEMS applications. *Microelectr. Eng.*, 77, 230–241, 2005.

53. Zubel, I., Silicon anisotropic etching in alkaline solutions III: On the possibility of spatial structures forming in the course of Si (100) anisotropic etching in KOH and KOH+IPA solutions. *Sens. Actuators*, 84, 116–125, 2006.
54. Wagner, A., KOH Si wet etch review, in: *Center for Nanoscale Science and Engineering*, North Dakota State University, lecture, 1–14, 2004.
55. Zubel, I. and Kramkowska, M., Etch rates and morphology of silicon (hkl) surfaces etched in KOH and KOH saturated with isopropanol solutions. *Sens. Actuators A*, 115, 549–556, 2004.
56. Tan, S.-S., Han, H., Reed, M.L., Process Induced defects on anisotropically etched Silicon. *Proc. Workshop on Micro Electro Mechanical Systems*, Japan, pp. 229–34, 1994.
57. Choi, W.K., Thong, J.T.L., Luo, P., Bai, Y., Tan, C.M., Chua, T.H., Formation of pyramids at surface of TMAH etched silicon. *Appl. Surf. Sci.*, 144-145, 472–475, 1999.
58. Fano, V., *Evaluación y estudio de procesos químicos anisotrópicos en la fabricación industrial de células solares*, PhD Thesis, University of the Basque Country (UPV/EHU). Spain, 2014.
59. Vallejo, B., Gonzalez-Mañas, M., Martinez-Lopez, J., Caballero, M.A., On the texturization of monocrystalline silicon with sodium carbonate solutions. *Sol. Energy*, 81, 5, 565–569, 2007.
60. Moynihan, M., O'Connor, C., Barr, B., Tiffaby, S., Braun, W., Allardyce GRentsch., J., Birmann, K., In-line and vertical texturing of mono-crystalline solar cells. *35th IEEE Photovoltaic Specialists Conference*, pp. 001028–001033, 2010.
61. Xi, Z., Yang, D., Que, D., Texturization of monocrystalline silicon with tribasic sodium phosphate. *Sol. Energy Mater. Sol. Cells*, 77, 255–263, 2003.
62. Campbell, S.A., Cooper, K., Dixon, L., Earwaker, R., Port, S.N., Schiffrin, D.J., Inhibition of pyramid formation in the etching of Si p<100> in aqueous potassium hydroxide-isopropanol. *J. Micromech. Microeng.*, 5, 209–218, 1995.
63. Kuchler, G., Müller, G., Brendel, R., Stereoscopic Reconstruction of randomly textured silicon surfaces. *Prog. Photovoltaics Res. Appl.*, 11, 2, 89–95, 2003.
64. Hylton, J.D., Kinderman, R., Burgers, A.R., Sinke, W.C., Bressers, P.M.M.C., Uniform pyramid formation on alkaline-etched polished monocrystalline (100) Silicon wafers. *Prog. Photovoltaics Res. Appl.*, 4, 6, 435–439, 1996.
65. Zubel, I., The influence of atomic configuration of (hkl) planes on adsorption processes associated with anisotropic etching of silicon. *Sens. Actuators A*, 94, 76–86, 2001.
66. Glembocki, O.J., Stahlbush, R.E., Tomkiewicz, M., Bias dependent etching of silicon in aqueous KOH. *J. Electrochem. Soc.*, 132, 1, 145–151, 1985.
67. Wefringhaus, E. and Helfricht, A., KOH/surfactant as an alternative to KOH/IPA for texturisation of monocrystalline silicon. *Proc. 24th EU PVSEC*, Hamburg, Germany.
68. Mayer, K., Kray, D., Orellana Perez, T., Schumann, M., Glunz, S.W., New surfactants for combined Cleaning and texturing of mono-crystalline

Silicon Waters alter wire-sawing. *23nd European Photovoltaic Solar Energy Conference and Exhibition*, Valencia, pp. 1109–1113, 2008.

69. Park, H., Lee, J.S., Lim, H.J., Kim, D., The effect of Tertiary-Butyl Alcohol on the texturing of crystalline silicon solar cells. *J. Korean Phys. Soc.*, 55, 1767–1771, 2009.
70. Silicon texture formulations with diol additives and methods of using the formulations, US8759231B2, Assigned to Intermolecular, Inc, United States, 2014.
71. Method for texturing silicon wafers for producing solar cells, US8182706B2, assigned to Universitat Konstanz, United States, 2009.
72. Kuchler, G. and Brendel, R., Reconstruction of the surface topography of randomly textured silicon. *Prog. Photovoltaic Res. Appl.*, 11, 89–95, 2003.
73. Cerofolini, G.F. and Meda, L., Chemistry at silicon crystalline surfaces. *Appl. Surf. Sci.*, 89, 351–360, 1995.
74. Fano, V., Rodriguez, V., Tejado, L., Jimeno, J.C., Towards in-line alkaline texturing for monocrystalline silicon solar cells. *Proceedings of 25nd European Photovoltaic Solar Energy Conference*, Valencia, Spain, 2010.
75. Choi, W.K., Thong, J.T.L., Luo, P., Tan, C.M., Chua, T.H., Bai, Y., Characterisation of pyramid formation arising from etching of silicon. *Sens. Actuators A*, 71, 238–243, 1998.
76. Cho, W.-J., Chin, W.-K., Kuo, Ch-T., Effects of alcoholic moderators on anisotropic etching of silicon in aqueous potassium hydroxide solutions. *Sens. Actuators A*, 116, 357–368, 2004.
77. Moldovan, C., Oosub, R., Dascalu, D., Nechifor, G., Anisotropic etching of silicon in a complexant redox alkaline system. *Sens. Actuators B*, 58, 438–449, 1999.
78. Pietsch, G.J., Higashi, G.S., Chabal, Y.J., Chemomechanical polishing of silicon: surface termination and mechanism of removal. *Appl. Phys. Lett.*, 64, 3115, 1994.
79. Zhu, X., Wang, L., Yang, D., Investigations of random pyramid texture on the surface of single-crystalline silicon for solar cells. *Proceedings of ISES Solar World Congress 2007: Solar Energy and Human Settlement*, pp. 1126–1130, 2007.
80. Lee, J., Lakshminarayan, N., Dhungel, S.K., Kim, K., Yi, J., Optimization of fabrication process of high-efficiency and low-cost crystalline silicon solar cell for industrial applications. *Sol. Energy Mater. Sol. Cells*, 93, 256–261, 2009.
81. Gangopadhyay, U., Dhungel, S.K., Mondal, A.K., Saha, H., Yi, J., Novel low-cost approach for removal of surface contamination before texturization of commercial monocrystalline silicon solar cells. *Sol. Energy Mater. Sol. Cells*, 91, 1147–1151, 2007.
82. Chan, R., O'Connor, C., Barr, R., Moynihan, M., Ridler, T., Allandyce, G., Wafer Cleaning and its effects on subsequent texturing processes. *24th European Photovoltaic Solar Energy Conference and Exhibition*, Hamburg, Germany, pp. 1199–1202, 2009.

83. Nussbaumer, H., Delhaye, F., Saule, W., Nagel, D., Queisser, S., Sander, B.-U., Kühnlein, H., Kösterke, N., Bay, N., Cimiotti, G., Eljaouhari, A., Schweckendiek, J., Dechant-Wagner, R., Sen, I., Kleinschmidt, S., Wet chemical processing in crystalline silicon wafer and cell production today and tomorrow. *24th Europen Photovoltaic Solar Energy Conference*, Hamburg, pp. 1356–1359, 2009.
84. Basu, P.K., Dhasmana, H., Udayakumar, N., Khan, K., Thakur, D.K., Regulated low cost pre-treatment step for surface texturing of large area industrial single crystalline silicon solar cell. *Solar Energy Materials & Solar Cells*, vol. 94, no. 6, pp. 1049–1054, 2010.
85. Angermann, H., Uredat, S., Zettler, J.T., Surface texturization and interface passivation of mono- and polycrystalline silicon substrates: Evaluation of wet chemical treatments by UV-NIR Reflectance. *24th European Photovoltaic Solar Energy Conference*, Hamburg, Germany, pp. 1954–1957, 2009.
86. Cao, Y., Liu, A., Li, H., Liu, Y., Qiao, F., Hu, Z., Sang, Y., Fabrication of silicon wafer with ultra low reflectance by chemical etching method. *Appl. Surf. Sci.*, 257, 7411–7414, 2011.
87. Krümberg, J., Melnyk, I., Schmidt, M., Michel, M., Fidler, T., Kagerer, M., Fath, P., Heilinger, L., Nussbaumer, H., New innovative alkaline texturing process for CZ silicon wafers. *24th European Photovoltaic Solar Energy Conference*, Hamburg, pp. 21–25, 2009.
88. Fano, V., Rodriguez, V., Freire, I., Azkona, N., Tejado, L., Jimeno, J.C., Progress in alkaline texturing inline for monocrystalline silicon solar cells. *Proceedings of 26nd European Photovoltaic Solar Energy Conference*, Hamburg, 2011.

3
Advanced Texturing

Lixia Yang*, **Yaoping Liu and Xiaolong Du**

Institute of Physics, Chinese Academy of Sciences, Beijing, China

Abstract

This chapter presents an overview of advanced texturing. First, metal-assisted chemical etching method is introduced as advanced texturing method. Second, the history and definition of metal-assisted chemical etching is introduced. Third, mechanisms of metal-assisted chemical etching in HF/metal salts/H_2O_2 are exhibited. It is an electrochemical reaction, and the driving force is the difference of the electrochemical potential between Silicon and redox systems. Fourth, methods and structure characteristics obtained by metal-assisted chemical etching methods are introduced. Two-step and one-step metal-assisted chemical etching methods are discussed, and the characteristics of topography and anti-reflection of the structures obtained using different metal catalysts and different etchant ratio are discussed. Fifth, metal-assisted chemical etching method is simple and low cost, and the obtained structures show good anti-reflection effect; thus, the industrialization and the application of this method in the mass production of Silicon solar cell are worth further researching.

Keywords: Advanced texturing, metal-assisted chemical etching, copper-assisted chemical etching, reflective losses, redox system, nanoparticles, one-step etching process, two-step etching process

3.1 Introduction to Advanced Texturing

Raw crystalline Silicon (Si) surfaces have a natural high reflectivity with a strong spectral dependence. The minimization of reflection losses is very important for high efficiency Si solar cells, and hence, surface texturing is applied as the first step for Si solar cells fabricating process. Anisotropic

**Corresponding author*: lxyang@iphy.ac.cn

Monika Freunek Müller (ed.) Photovoltaic Manufacturing: Etching, Texturing, and Cleaning, (83–114) © 2021 Scrivener Publishing LLC

alkaline texturing usually is performed for monocrystalline-Silicon (c-Si) to obtain pyramid structure [1–5], while isotropic acidic texturing typically is used for multicrystalline-Silicon (m-Si) to obtain "worm-like" structure [6]. These structures, known as anti-reflection structures, can reduce the reflectivity by offering the incident light a secondary chance to reenter the Si substrate [7–10]. Moreover, these texturing methods, known as conventional texturing methods, have been industrialized. In recent decades, metal-assisted chemical etching method [11–19] has been developed and has attracted extensive studies in the application for further reducing the reflection losses for Si solar cells. Metal-assisted chemical etching method generally carries out at room temperature for several minutes, which greatly reduces the energy consumption in comparison with the conventional alkaline texturing method, which takes tens of minutes at a relatively high temperature. In addition, compared to the microscopic-level pyramid structures for c-Si and the microscopic-level worm-like structures for m-Si achieved by conventional texturing methods, nanowires [20–23] and nanopores [21, 23] are usually obtained by utilizing metal-assisted chemical etching methods. These nanostructures provide a gradually varying refractive index. Because the refractive index of the nanostructures layer is between the refractive index of air and Si, they can suppress the reflection over a broad spectral range [24–31]. As a result, "black" Si (B-Si) anti-reflection layer is formed. In contrast to conventional texturing methods, metal-assisted chemical etching method is called advanced texturing method.

In this chapter, a detailed introduction to advanced texturing method is exhibited, including history and definition (Section 3.2), mechanisms (Section 3.3), and methods and structure characteristics obtained by advanced texturing method (Sections 3.4 and 3.5).

3.2 History and Definition of Metal-Assisted Chemical Etching

Etching Si in aqueous solutions of hydrofluoric acid (HF) containing an oxidizing agent, such as nitric acid (HNO_3), was first studied more than 50 years ago [32, 33]. Researchers showed that nanostructures could be formed on the Si surface. This method of Si etching, which uses high dilution of HF and oxidant, is commonly referred to as "stain etching". In 1997, stain etching further evolved by the discovery that the etching rate can be accelerated by adding a metal catalyst [34]. Porous Si was created using a catalyst of deposited Al film on the Si substrate submerged in an etching solution of $HF/HNO_3/H_2O$. The reaction between Al and HNO_3 produced the required charges for the fast initiation of chemical etching, which dramatically reduced the incubation time necessary for the formation of porous Si [34].This is the first demonstration of metal-assisted chemical etching. Shortly after, a simpler variant was investigated by Li and Bohn, who found that a thin layer (nominal thickness <10 nm) of Au, Pt, or Au/Pd deposited on the Si (100) surface prior to immersion in aqueous HF and hydrogen peroxide (H_2O_2) facilitated the etching, which led to a simple and effective way of producing porous Si with different morphologies and light-emitting properties [35]. These etching methods described by Li and Bohn inspired various approaches to fabricate Si-based nanostructures.

A typical metal-assisted chemical etching process usually includes two steps: metal deposition and electrochemical etching. The metal nanoparticles, used as catalysts for the etching of Si, are deposited on Si wafers by various techniques such as sputtering [35], thermal evaporation [24, 36, 37], electrode plating [38], and electroless plating [23, 39–42]. The Si wafers, covered with metal nanoparticles, are next immersed in aqueous etching solution containing HF and oxidants. Si in contact with the metal nanoparticles is etched much faster than the area without the metal nanoparticles coverage; thus, shallow pits are formed under these metal nanoparticles. As a result, the metal nanoparticles fall into these pits, catalyzing the formation of nanopores [21, 23, 43–46] or nanowires [20–23, 47–52] in the Si substrate.

To simplify the metal-assisted chemical etching process, a one-step metal-assisted chemical etching method has been developed where the metal deposition and the electrochemical etching occur at the same time. The topology of the Si structures is significantly affected by the etching condition and the deposition method and morphologies of the noble metals used as catalysts, plus the composition of the solution, all of which will be discussed in the following sections.

3.3 Mechanisms of Metal-Assisted Chemical Etching

3.3.1 Reactions of Metal-Assisted Chemical Etching

Metal-assisted chemical etching mainly contains three regimes: 1. Si covered with metal nanoparticles immersed in HF/oxidants (not including metal salts) solution; 2. Bare Si immersed in HF/metal salts solution; 3. Bare Si immersed in HF/metal salts/oxidants solutions. Regime 1 is two-step metal-assisted chemical etching method, while regime 2 and 3 is one-step metal-assisted chemical etching method, the mainly difference between regime 2 and regime 3 is the concentration of metal salts, the concentration is usually high in regime 2, which will be presented in the following section. The metal-assisted chemical etching reactions and mechanisms of these regimes are essentially the same; hence, regime 3 is used to illuminate the reactions and mechanisms of metal-assisted chemical etching. Because different oxidants show similar action in the metal-assisted chemical etching process, only the mechanism in solution of HF/metal salts/H_2O_2 will be discussed. The corresponding reaction consists of two half-cell reactions: a cathode reaction at the metal nanoparticles surface and an anode reaction occurring at the contact area between the Si and metal nanoparticles.

The possible cathode reaction could be written as:

$$M^{n+} + ne^{-} \rightarrow M^{0}(s) \tag{3.1}$$

$$H_2O_2 + 2H^{+} + 2e^{-} \rightarrow 2H_2O, \tag{3.2}$$

where M is a metal and n is valence.

Meanwhile, it is still controversial whether the reduction of H^+ into H_2 occurs at the cathode:

$$2H^{+} \rightarrow H_2\uparrow + 2h^{+} \tag{3.3}$$

There are theories where H_2 is formed: Li and Bohn [35] and Harade [53] *et al.* proposed that the reduction of H^+ is another cathodic reaction (Equation 3.3); while Chartier *et al.* [54] proposed the gas came from the anodic reaction, because there is no gas bubbling and no obvious etching occurring when the Pt covered Si substrate is immersed in the HF solution for 24 hours [38]. Moreover, the possibility that gas generated from the dismutation of H_2O_2 is excluded, because the gas does not occur on an Ag-loaded Si substrate subjected to high concentrations of H_2O_2 solution without HF [54].

At the anode, the Si is oxidized and dissolved simultaneously. There are theories about the dissolution of Si, which can be catalogued into three groups: [11] i) direct dissolution of Si in the tetravalent state; [17, 19, 20, 53, 55, 56] ii) direct dissolution of Si in the divalent state; [54, 57] and iii) Si first oxidized into SiO_2 and then subsequently dissolved [41, 58–62]. Because it is difficult to perform *in situ* exploration of the surface state and the uncertainty of *ex situ* studies of the Si surface state, the electrochemistry of the Si dissolution remains an open question.

Chartier *et al.* summarized the mechanisms proposed by many researchers, a generic anodic reaction: [54]

$$Si + 6HF + mh^{+} \rightarrow H_2SiF_6 + mH^{+} + [(4\text{-}m)/2]H_2\uparrow \tag{3.4}$$

m is the number of holes per dissolved Si atom; ranging from 2 to 4. Taking into account the cathodic reduction of H_2O_2, the overall reaction is:

$$Si + 6HF + [m/2]H_2O_2 \rightarrow H_2SiF_6 + mH_2O + [(4\text{-}m)/2]H_2\uparrow \quad (3.5)$$

3.3.2 Redox System of Metal-Assisted Chemical Etching

The generally accepted theory is that metal-assisted chemical etching occurs as a localized electrochemical process [63–65], in which the reduction of oxidant and the oxidation and dissolution of Si occur simultaneously at the Si surface, with the charges being exchanged between the oxidant and Si through the metal nanoparticles. Holes injection is the charge transfer process for metal-assisted chemical etching of Si [20, 66, 67], with the driving force of the holes injection being the difference of electrochemical potential between the Si and redox species [11, 19, 23]. Figure 3.1 shows the energy levels of the redox species, that lie well below the Si valence band E_v; thus, the electrochemical reaction on Si with M^{n+} and H_2O_2 is expected to occur by injection of holes (*h*) into the valence band of Si. Although H_2O_2 is expected to inject holes into the valence band of Si from the energy point; bare Si subjected to HF/H_2O_2 solution produce no obvious topological features. Control experiments have been performed, in which Si wafers with and without metal coatings were subjected to the HF/H_2O_2 solution for the same period of time, Si wafers without metal coatings produced no obvious topological features, whereas Si wafers covered with metal coatings

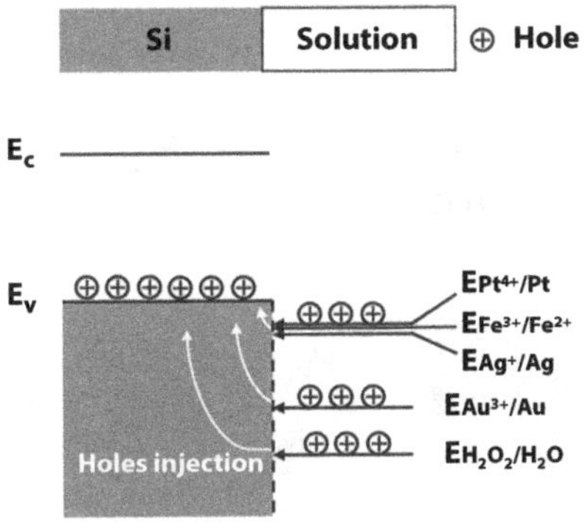

Figure 3.1 Schematic of the potential relationship between bands in a Si substrate, where E_c and E_v are the conduction and valence bands, respectively, and standard potentials of various oxidants. Adapted with permission [11]. Copyright 2011, Wiley-VCH. Adapted with permission [23]. Copyright 2006, Wiley-VCH.

generated topographically rough porous Si [35]. These results indicate that the metal is necessary for the fast reduction of H_2O_2 and H_2O_2 is reduced on the metal surface, not at the Si surface [11].

A schematic describing the sequence of the processes in the metal-assisted chemical etching technique is shown in Figure 3.2 [11]. The Si/HF/metal salts/H_2O_2 system is proposed to be composed of a corrosion-type redox couple: the cathodic reduction of M^{n+} and H_2O_2 and at the same time, the anodic oxidation and dissolution of SiO_2, which occur locally beneath the metal nanoparticles [19]. The sequence is as follows [11].

i. M^{n+} ions withdraw electrons from the Si substrate and metal nanoparticles are deposited on the Si surface; generally, charge exchange between Si and M^{n+} is more likely to take place at the steps and other surface defects [23, 68]. As the metal nanoparticles are more electronegative than Si, they will further capture electrons from the Si substrate and become negatively charged. Therefore, other M^{n+} ions will come to the negatively charged metal nanoparticles, receive electrons from the them, and are then reduced to metal nanoparticles and adsorb there, leading to the growth of more metal nanoparticles. Meantime, the H_2O_2 injects holes to the Si substrate preferentially through the metal nanoparticles and is reduced;
ii. The injected holes move to the Si substrate beneath the metal nanoparticles;
iii. The Si underneath the metal nanoparticles releases electrons as required by M^{n+} ions and H_2O_2 to be reduced and, in turn,

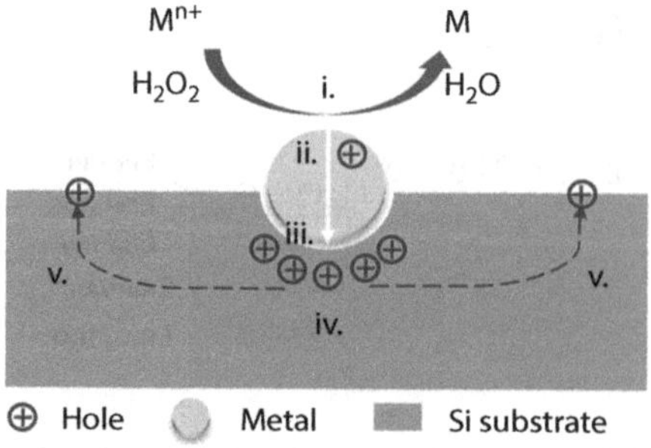

Figure 3.2 Schematic of the electrochemical processes involved in metal-assisted chemical etching. The numbers indicate the process step sequences introduced in Section 3.2. Adapted with permission [11]. Copyright 2011, Wiley-VCH.

is oxidized by the injected holes and etched by HF. Shallow pits form under the metal nanoparticles and then the metal nanoparticles fall into the pits;

iv. The concentration of injected holes is at their maximum at the Si/metal interface; thus, the etching is faster at the metal covered area of the Si substrate than the bare Si substrate;

v. The unconsumed holes at the Si/metal nanoparticles interface will diffuse to other areas and induce etching if the rate of holes consumption is lower than the rate of holes injection.

In the metal-assisted chemical etching process, it is worth noting that lateral etching does not occur. With the reduction of M^{n+} ions and H_2O_2, and the dissolution of Si, a charge-depletion layer with high resistance emerges around the metal nanoparticles [23], as shown in Figure 3.3. Hence, the metal nanoparticles site, where electrons are consumed, and the Si site, where electrons are released, should not be far apart for the sake of realizing the smooth transport of electrons [23, 69]. Researchers suggest that the charge exchange and transport between the anodic and cathodic sites is more favorable at the bottom of the pore (metal/Si interface) than at the pore wall, because the bottom has the shortest electrons transport distance, and because of the unique catalytic property of the metal nanoparticles, which greatly decreases the activation energy for the cathodic reaction [23]. That is, longitudinal etching would be more favored than lateral etching; thus, deep pores or long wires would be formed, owing to the prolonged longitudinal etching of Si [23]. The rate of penetration, R_p, of the

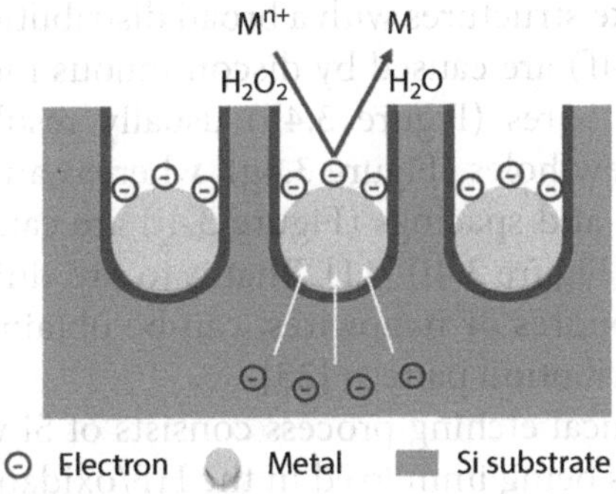

Figure 3.3 Mechanism of hindering the lateral dissolution of Si along the pore during metal-assisted chemical etching in HF solution. Adapted with permission [23]. Copyright 2006, Wiley-VCH.

metal nanoparticles into the bulk Si is dependent on the depth of pores or the length of the wires in the Si substrate, *D*, and is linearly proportional to the etching time, *t*, where Δm represents the total mass loss of the Si wafer; *S* is the area of Si, and *d* is the density of Si [54].

$$D = \frac{1}{2}\frac{m}{dS}; \qquad R_p = \frac{D}{t} \tag{3.6}$$

The above proposed mechanism of metal-assisted chemical etching is generally applicable. However, it may appear different when come to different metal catalysts, different oxidants and different etchants' concentrations. In regime 1, metal nanoparticles are deposited on the Si wafers before being immersed in the HF/oxidants solution, then the holes injection is supplied by the oxidants through the metal nanoparticles. In regime 2, holes injection is fully completed by the metal salts.

3.4 Methods of Metal-Assisted Chemical Etching

3.4.1 Two-Step Metal-Assisted Chemical Etching Process

The morphologies of the metal-assisted chemical etching structures are strongly influenced by the morphologies of the metal nanoparticles. Figure 3.4 shows representative morphologies of the metal catalysts and the corresponding morphologies of the etched structures [11]. Well-defined pores (Figure 3.4b) usually result from well-separated metal nanoparticles (Figure 3.4a); while wall-like or wire-like structures (Figure 3.4d) are caused by more compactly distributed metal nanoparticles (Figure 3.4c); wall-like or wire-like structures with a broad distribution of diameters and spacings (Figure 3.4f) are caused by discontinuous metal patches (Figure 3.4e); well-defined wires (Figure 3.4h) usually result from continuous metal film with a few holes (Figure 3.4g); whereas an array of wires with identical diameters and spacings (Figure 3.4j) are caused by a metal film with regular holes (Figure 3.4i) [11]. That is to say, different Si nanostructures, such as nanopores or nanowires, can be obtained by adjusting the metal catalysts distribution pattern [23].

The electrochemical etching process consists of Si wafers covered with metal nanoparticles being immersed in the HF/oxidants (e.g., H_2O_2) solution. The concentrations of HF and H_2O_2 not only affect the etching rate, but also the morphologies of the etched structures. Chartier *et al.* systematically investigated the influence of the molar ratio $\rho = [HF]/([HF]+[H_2O_2])$

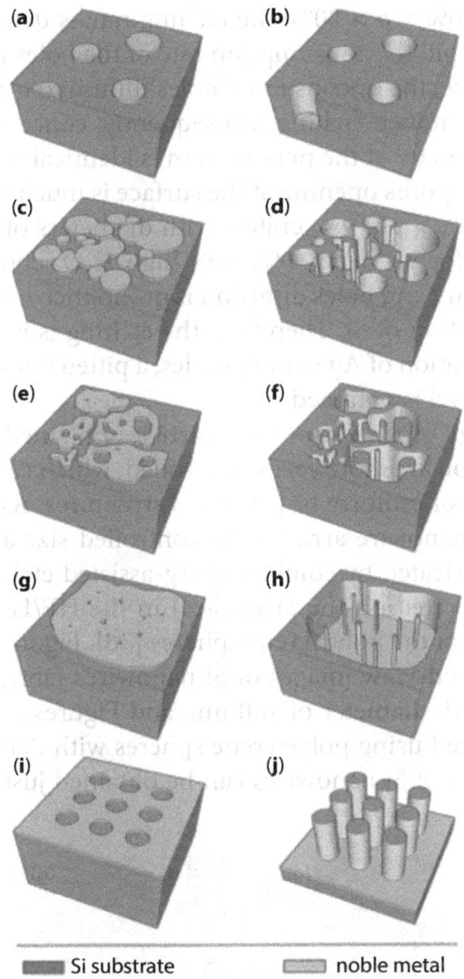

Figure 3.4 Schematic of morphologies of etched structures (right column) induced by metal catalysts with differently shaped noble metals (left column). Adapted with permission [11]. Copyright 2011, Wiley-VCH.

on the etching rate and on the post-etched morphologies [54]. The etching rate reaches its maximum when ρ is ~85%, while varying ρ about this value leads to a decrease of the etching rate. In addition, the morphologies of the resulted structures are determined by ρ. For 100% > ρ > 70%, the etching rate is determined by [H_2O_2] and nearly all injected holes are consumed at the Ag/Si interface at the pore bottom because [HF] is high enough to dissolve SiO_2, as a result, straight or curved cylindrical pores with a diameter matching the size of the Ag nanoparticles at their bottom are formed by the

Ag catalysts; for 70% > ρ > 20%, the etching rate is determined by [HF]. In this case, at the pit, the consumption rate of the holes is slower than the injected rate. Hence, the unconsumed holes diffuse to other regions (such as pore wall) and induce etching; consequently, coned-shaped pores are produced. The diameter at the pore bottom is identical to the Ag nanoparticle size, while the pore's opening at the surface is much larger than that of the bottom; for 20% > ρ > 9%, craters with diameters of several micrometers are formed; for 9% > ρ > 0%, very high H_2O_2 concentration, holes injection is abundant and holes diffusion is pronounced, leading to the oxidation and dissolution of Si. Therefore, the etching is isotropic and independent on the location of Ag nanoparticles; a pitted but smooth surface at the microscopic level is obtained [54].

The morphology of the metal nanoparticles obtained by sputtering or thermal evaporation are more easily controlled, which can be used to catalyze the formation of uniform or patterned structures. As shown in Figure 3.5, large-area Si nanowire arrays with controlled size and packing density have been fabricated by combining Ag-assisted etching (Ag film that is thermally evaporated on the Si surface) in the HF/H_2O_2 solution with the masking effect of the polystyrene spheres [36]. Figure 3.5a show plane-view and b 15° tilted-view images of Si nanowires fabricated using polystyrene spheres with diameter of 100 nm, and Figures 3.5c and d show Si nanowires fabricated using polystyrene spheres with diameter of 180 nm. Different diameters of Si nanowires can be obtained just by adjusting the

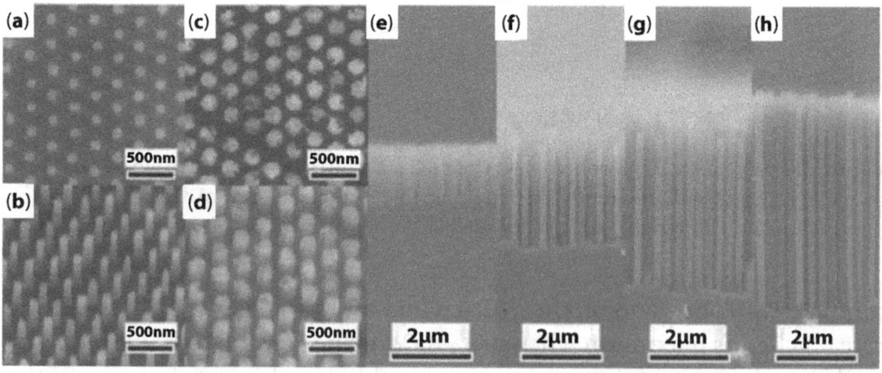

Figure 3.5 Scanning electron microscope images of Si nanostructures with polystyrene spheres at a nominal diameter of 260 nm. Plane-view and tilted-view (~15°) images of samples fabricated using polystyrene spheres with a diameter of (a, b) 100 nm and (c, d) 180 nm. Cross-sectional SEM images of Si nanostructures after etching for (e) 4 min, (f) 8 min, (g) 12 min, and (h) 16 min. Adapted with permission [36]. Copyright 2007, Wiley-VCH.

diameter of the polystyrene spheres [36]. Figures 3.5e–h show cross-view images of Si nanowires after etching for 4, 8, 12, and 20 min, respectively. These results show that the length of the nanowires can be controlled by varying the etching time [36]. Therefore, patterned structures with controlled diameter, length, and density can be obtained by combining the two-step metal-assisted chemical etching technique with the polystyrene spheres lithography method [36].

However, the physical deposition methods are costly and time-consuming, restricting widespread application of the metal-assisted chemical etching process. In contrast, electroless deposition is low cost and simple for the deposition of metal nanoparticles. The method is widely utilized to deposit metal nanoparticles as catalysts if there are no strict requirements (such as the controlled diameter of the nanostructures and the controlled spacing between the nanostructures) for the obtained structures. Consequently, metal nanoparticles are randomly distributed, resulting in the formation of random nanostructures [62]. Figures 3.6a–c show random pores induced by depositing low density Ag nanoparticles, the subsequently etched with a HF/H_2O_2 solution. The silver nanoparticles can be observed at the end of the pores [62]. Note that the Ag nanoparticles are deposited in a $HF/AgNO_3$ solution. In comparison, aligned Si nanowires are produced using Ag nanoparticles with high density (Figures 3.6d–f) [62]. The density of Ag nanoparticles is controlled during the deposition process in a $HF/AgNO_3$ solution, and thus, different nanostructures can be produced [20]. This is similar to the cases found in Au-assisted chemical etching [21–23]. However, in the case of Pt-assisted chemical etching of Si, either shallow pits [20], straight nanopores [56], or curvy pores [23, 39, 70] are formed. The behavior of Pt nanoparticles is complex, and it is not clear that the intrinsic difference in the catalytic activity between the Ag and Pt up to now. In addition, only shallow pits are produced when Cu-covered Si is immersed in the HF/oxidants solution [20, 23, 39]. The reason may be the Cu nanoparticles are dissolved before they can catalyze the etching of Si [23, 39]. A detailed study on Cu-assisted chemical etching will be presented in the following section.

3.4.2 One-Step Metal-Assisted Chemical Etching

In the two-step metal-assisted chemical etching method, metal deposition generally requires complex and expensive equipment, performed at high temperature and under high vacuum. These attributes make the process time-consuming and expensive. Besides, black Si fabrication over a large area is difficult to achieve with the two-step metal-assisted

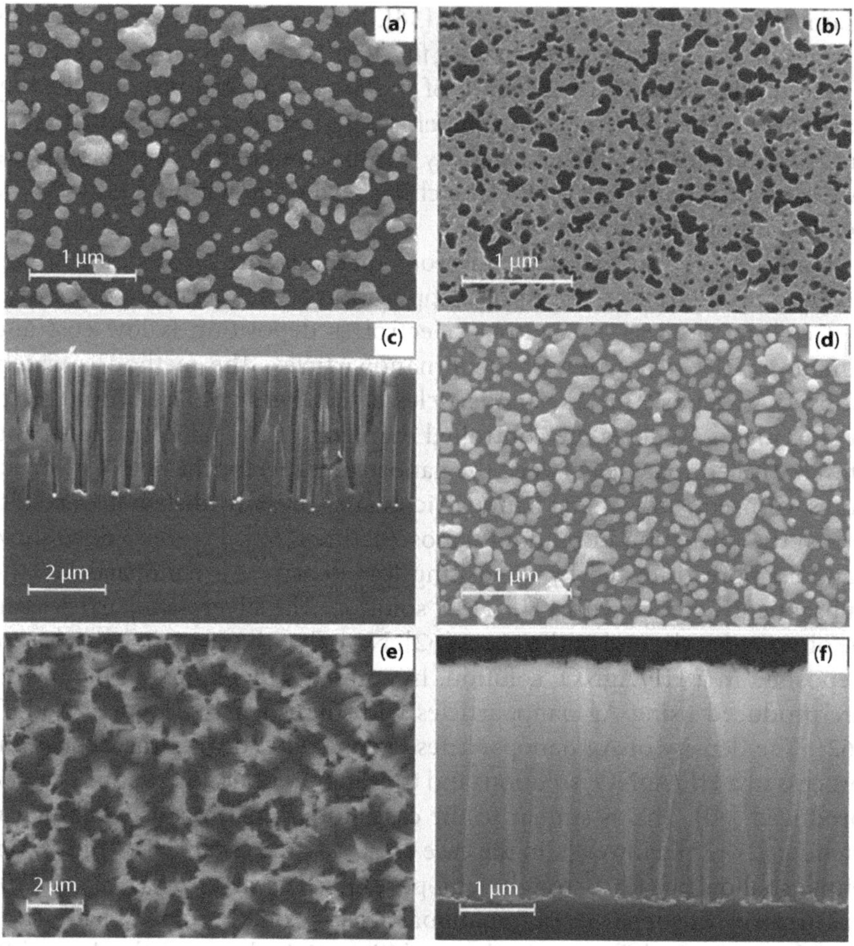

Figure 3.6 Scanning electron microscope investigation of Si nanostructures induced by Ag nanoparticles with different densities on a Si (100) substrate in HF/H_2O_2 solution; (a) Ag nanoparticles on the Si surface after 5-min ultrasonication, (b, c) top and cross-sectional views of the Ag-nanoparticle-induced Si nanostructures with long pores after 5 min of etching in HF/H_2O_2 solution, (d) SEM images of Ag nanoparticles on a Si surface without ultrasonication, and (e, f) top and cross-sectional views of the Ag-nanoparticle-induced Si nanowire arrays after 5 min of etching in HF/H_2O_2 solution. Adapted with permission [62]. Copyright 2008, Wiley-VCH.

chemical etching method. Consequently, the two-step metal-assisted chemical etching method evolved into a one-step metal-assisted chemical etching method, which the metal deposition and the electroless chemical etching occur simultaneously. Branz *et al.* developed a one-step Au-assisted chemical etching method to fabricate nanoporous B-Si with the etchants consisting of $HAuCl_4/HF/H_2O_2$ [71]. The formation of Au nanoparticles resulting from the reduction of $HAuCl_4$ is indicated by the observation of Au nanoparticles at the ends of nanopores, as shown in the inset of Figure 3.7a, confirming the local catalytic action of an Au nanoparticle [71]. The reduction of H_2O_2 at the Au nanoparticles causes holes injection into the Si substrate, resulting in the rapid local oxidation of Si, which is then etched by HF forming densely distributed nanopores near the surface, as shown in Figure 3.7a [71]. In addition, the nanopores made the near-surface optical constants vary gradually, resulting in the reduced reflectance, as shown in Figure 3.7b. Branz *et al.* compared the total reflectance spectra of a polished Si (100) to identical wafers etched in the 5-nm colloidal Au solution for 900 s and etched in the $HAuCl_4$ solution for 20~180 s, which show that with the longer etching time in the $HAuCl_4$ solution, the reflectance is lowered to 2%; the sample etched with colloidal Au is below about 8% from 300 to 1,000 nm. In addition, the reflectance of B-Si rises at $\lambda > 1{,}000$ nm due to the light reflected from the wafer back surface and sample holder [71].

The cost of $AgNO_3$ is much less than that of $HAuCl_4$, and thus, $AgNO_3$ is promising for commercial fabrication of B-Si and for mass production of B-Si based solar cells [72, 73]. Figures 3.8a and b show scanning electron microscope (SEM) images of the p-type Si (p-Si) (100) surface after being etched in HF/$AgNO_3$ solution for 30 s before and after removing the

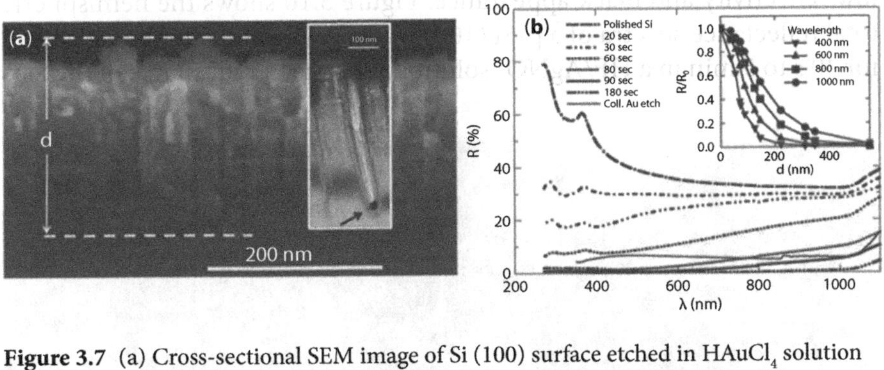

Figure 3.7 (a) Cross-sectional SEM image of Si (100) surface etched in $HAuCl_4$ solution for 80 s, with dashed lines indicating extrema of the density-grade depth found by computerized pixel correlation analysis. Inset to (a) is dark-field TEM (100-nm scale indicated) of a nanopore and an Au nanoparticle (indicated by arrow) produced by $HAuCl_4$ solution etching, (b) Reflectance spectra taken after etching in $HAuCl_4$ black Si-etch solution for 80 s [SEM of (a)] and other times, as indicated; additional spectrum taken after 900 s etch in colloidal Au solution. Inset shows R at indicated wavelengths, normalized to the polished wafer (d=0), vs. density-grade depth, d. Adapted with permission [71]. Copyright 2009, AIP Publishing LLC.

Ag nanoparticles by HNO_3 [74]. The size of these Ag nanoparticles ranges from 10 to 100 nm, which are densely embedded in the bulk Si. An abundance of nanopores can be observed on the Si surface after removing the Ag nanoparticles (metal nanoparticles can be removed by ultrasonic in the concentrated HNO_3 solution), with the nanopores' diameter being similar to that of the Ag nanoparticles [74]. The Si/$AgNO_3$/HF system is composed of a corrosion-type redox couple. At the start of the reaction, the Ag^+ ions in the vicinity of the Si surface withdraw electrons from the Si, and then, the Ag nanoparticles are deposited on the Si surface; Simultaneously, the Si under the Ag nanoparticles is oxidized to SiO_2 because the Si releases as many electrons as required by the Ag^+ ions [74]. Shallow nanopores form immediately under the Ag nanoparticles due to the etching of SiO_2 by the HF. As a result, Ag nanoparticles sink into the nanopores, which is confirmed by the SEM images of the etched Si surface before and after the Ag nanoparticles are removed [74], exhibited in Figures 3.8a and b. With longer etching time in the HF/$AgNO_3$ solution, the Ag nanoparticles that do not enter the pores grow into branched Ag dendrites, and the Ag nanoparticles that enter the pores sink deeper and deeper into the bulk Si. Figures 3.9a and b show cross-view SEM images of the p-Si (100) surface etched in the HF/$AgNO_3$ solution for 4 and 5 min, after the Ag nanoparticles are removed. Uniform nanowires are obtained. The depth of the nanowires is ~200 nm for 4 min processing and ~250 nm for 5 min processing [74]. The Si nanowires exhibit a gradually varying refractive index, which lead to the low reflectivity and black appearance. Figure 3.10 shows the hemispheric total reflectance spectra of a p-Si (100) wafer before and after being etched for 30 s to 5 min in a HF/$AgNO_3$ solution. The mean reflectance of the Si is

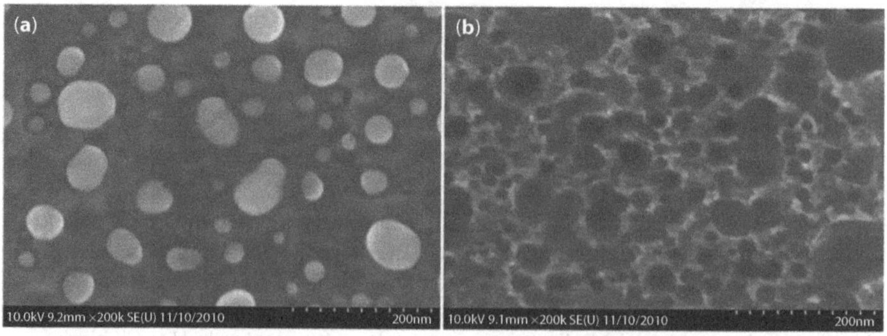

Figure 3.8 (a) Scanning electron microscope image of Ag nanoparticles on Si etched by HF/$AgNO_3$ solution for 30 s. (b) Shallow nanopores formed in the top surface of Si substrate after the Ag is removed. Adapted with permission [74]. Copyright 2012, Wiley-VCH.

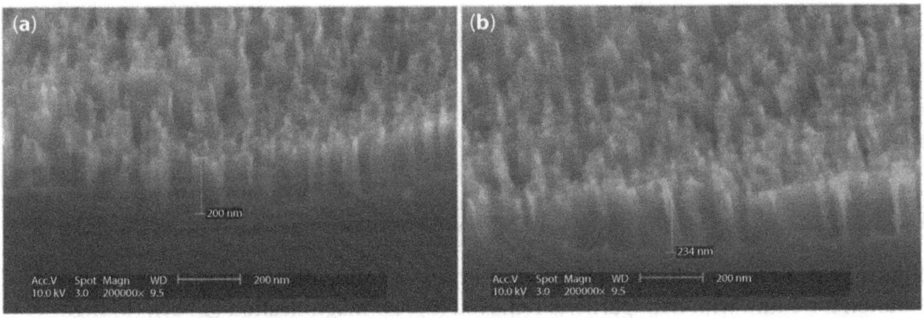

Figure 3.9 Cross-sectional SEM images of p-Si (100) surfaces, after Ag is removed, after etching in HF/$AgNO_3$ solution for (a) 4 min and (b) 5 min. Adapted with permission [74]. Copyright 2012, Wiley-VCH.

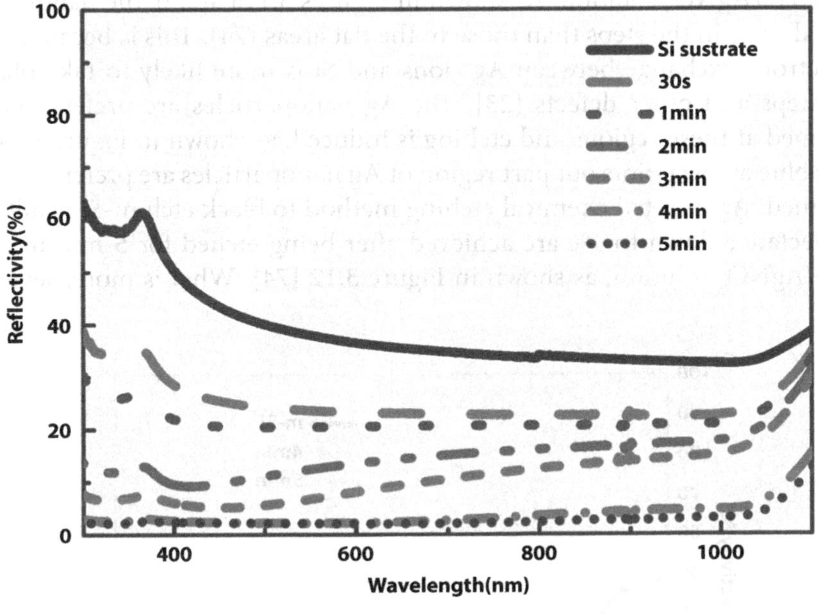

Figure 3.10 Hemispheric total reflectance of the p-Si (100) wafer, after being etched in HF/$AgNO_3$ solution from 30 s to 5 min, compared to an unetched surface. Adapted with permission [74]. Copyright 2012, Wiley-VCH.

significantly reduced for longer etching times from 30% to 2% at the wavelengths range from 300 to 1,000 nm [74].

This technique can also be applied to multi-crystalline Si (m-Si) [74]. Figure 3.11 shows the SEM images of m-Si after etching in HF/$AgNO_3$ solution for 5 min. The m-Si is first etched in a HNO_3/HF/CH_3COOH solution

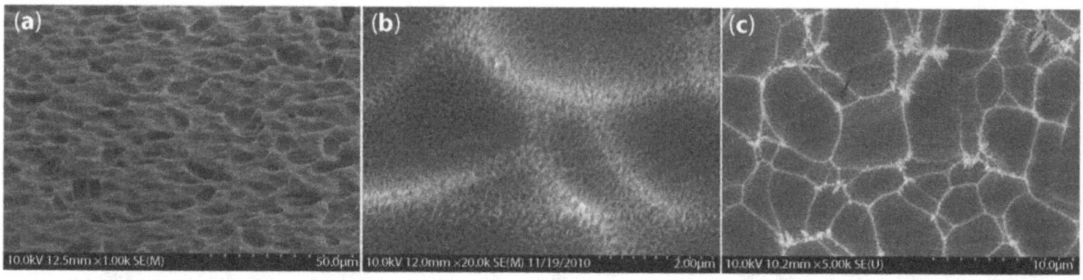

Figure 3.11 Scanning electron microscope images of m-Si after etching in HF/$AgNO_3$ solution for 5 min with nanoparticles removed (a) low magnification (1 KX), (b) high magnification (20 KX) (c) after etching in HF/$AgNO_3$ solution for 30 s without removing the Ag nanoparticles (5 KX). Adapted with permission [74]. Copyright 2012, Wiley-VCH.

(volume ratio 10:2:5) for 90 s to remove saw damage, then is black etched in a HF/$AgNO_3$ solution. As shown in Figures 3.11a and b, the nanowires are deeper in the steps than those in the flat areas [74]. This is because the electrons exchange between Ag^+ ions and Si is more likely to take place at steps and other defects [23]. The Ag nanoparticles are preferentially formed at these regions and etching is induced, as shown in Figure 3.11c, the blue arrow points out part region of Ag nanoparticles are preferentially formed. Ag-assisted chemical etching method to black etch m-Si result in reflectance down to 4% are achieved after being etched for 5 min in the HF/$AgNO_3$ solution, as shown in Figure 3.12 [74]. What is more, several

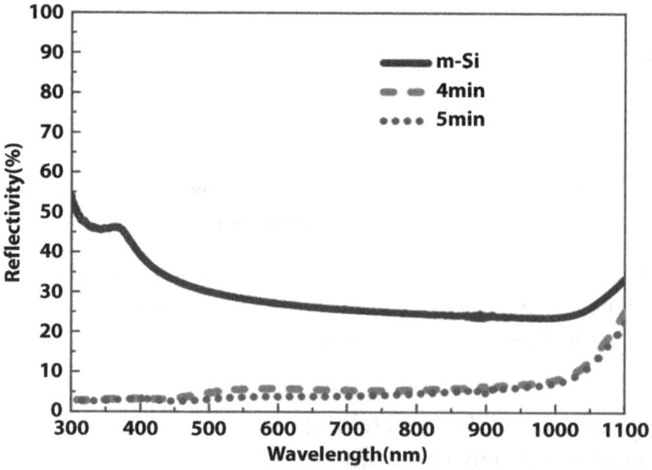

Figure 3.12 Hemispheric total reflectance of a m-Si wafer; unetched and after etching in HF/$AgNO_3$ solution for 4 and 5 min. Adapted with permission [74]. Copyright 2012, Wiley-VCH.

solar cells were fabricated and tested based on the Ag-assisted chemical etched 156 mm × 156 mm m-Si in Tianwei New Energy Holdings Co [74]. At first, the m-Si wafer is etched in HF/$AgNO_3$ solution for 3 min after removing the saw damage. In addition, 80-nm SiNx passivation film is directly plasma enhanced chemical vapor deposited on the phosphorous diffused m-Si wafer emitter with square resistivity of 65 Ω/sq, and then, the standard Al back-surface field and screen printed Ag-based front grid are applied to the m-Si wafer [74]. The results measured by using Halm cetisPV-CT-L1 equipment with calibrated 1 Sun simulators shows that the conversion efficiency of the m-Si solar cell is 14.9%, with an open-circuit voltage of 596 mV, short-circuit current of 32.42 mA/cm^2 and fill factor of 77.1% [74]. Their further study reveals that the thermal oxidation followed with the deposition of SiNx passivation film is very effective to passivate the nanostructures of the black Si. The thermal oxidation is performed on the 3 min Ag-assisted chemical etched m-Si wafer under oxygen atmosphere at 800°C for 20 min, and then, 60-nm SiNx is deposited on this oxidized surface. The conversion efficiency of the cell is increased to 15.8%, with open-circuit voltage of 604 mV, short-circuit current of 33.89 mA/cm^2 and fill factor of 77.3% [74]. The increase of performance is attributed to the efficient oxidation passivation of the nanostructure, because it is difficult for SiNx to passivate the bottom of the nanostructures; therefore, they conclude that the passivation plays a key role in the fabrication of black Si solar cells with high conversion efficiency [74]. Hence, it is important for B-Si solar cell to find a suitable depth of the nanopores, which not only has a low reflection loss, but also has a low recombination loss.

Oxidants, such as H_2O_2 or $Fe(NO_3)_3$, are not necessary for the one-step metal-assisted chemical etching process. In other words, metal-assisted chemical etching can proceed both in HF/metal salts/oxidants solutions (regime 3) and HF/metal salts solutions (regime 2). By controlling the concentration of the etchants, the etching time, B-Si with different morphologies and reflectance can be achieved.

3.5 Copper-Assisted Chemical Etching

3.5.1 Copper-Assisted Chemical Etching of c-Si

Au and Ag are the most frequently used noble metals for the metal-assisted chemical etching method; however, their high cost restricts them from large-scale applications. Thus, finding a low-cost metal precursor for the metal-assisted chemical etching process is attracting more attention of researchers. In early studies, Cu was utilized to replace the expensive noble metal catalysts for fabricating B-Si [67, 75].However, instead of the desirable nanopores or nanowires, only shallow pits [20, 23, 69] were formed, limiting the effectiveness of the surface as an anti-reflection layer. Figure 3.13a shows the SEM image of Cu nanoparticles formed on the Si surface in the HF/$Cu(NO_3)_2$ solution; compact metal grain films are formed, which restrict the etching of SiO_2 by HF due to the mask effect of these Cu films [23]. Therefore, an oxidant is necessary for oxidizing the Cu to Cu^{2+} ions and then suppressing the formation of Cu films. However, only an abundance of shallow pits are observed after the treatment in the HF/$Fe(NO_3)_3$ solution, as shown in Figure 3.13b, and no Cu nanoparticles are observed [23]. This is because the electrochemical potential of Cu/Cu^{2+} is lower than that of Fe^{3+}/Fe^{2+}; thus, Fe^{3+} will inject holes into Cu, causing the oxidation of Cu to Cu^{2+} ions [23]. As the Cu films usually are dissolved in the HF/oxidants solution, continuous catalyst behavior is not sustainable, thus B-Si cannot be obtained. Researchers concluded that there are two reasons for the unsuccessful etching of Si using Cu as the catalyst: 1) Cu nanoparticles tend to form dense films on the Si surface which act as a protection layer, making it difficult to etch Si [20–22]; and 2) Cu nanoparticles dissolve in solutions containing oxidants [such as H_2O_2 or $Fe(NO_3)_3$], making it difficult to etch Si without Cu as the catalyst [23].

Hierarchically rough Si surfaces with micro and nanostructures have been successfully fabricated by Lee *et al. via* immersing Cu nanoparticles plated Si wafer in a HF/H_2O_2 solution for 2 hours [76]. By introducing H_3PO_3 to the etching solution as a reducing agent and extending the etching time to 8 hours, Lu *et al.* synthesized inverted pyramid nanoporous

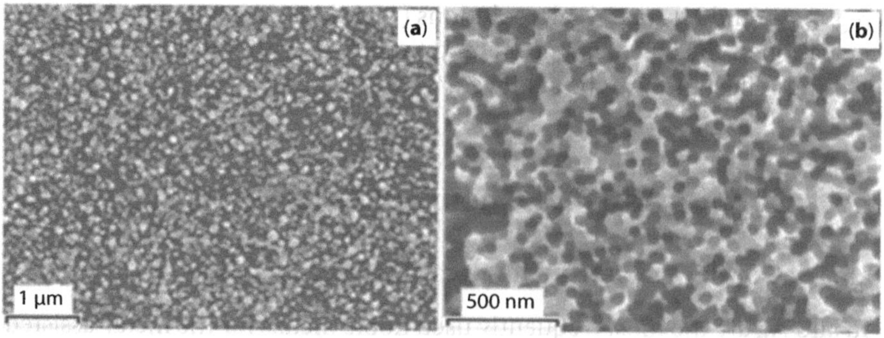

Figure 3.13 (a) Scanning electron microscope image of Cu nanoparticles deposited on Si from HF/$Cu(NO_3)_2$ solution. (b) Shallow pits formed on a Cu-treated Si substrate. Adapted with permission [23]. Copyright 2006, Wiley-VCH.

B-Si with mean reflectivity of 0.96% using a one-step Cu-assisted chemical etching technique at room temperature [67]. Although B-Si can be obtained by this Cu-assisted chemical etched technique, it takes too much time, inhibiting its widely application in the fabricating of B-Si. Uniform nanopores with diameter of 1 μm on the top and tapering to hundreds of nm at the base was achieved by Yu *et al.* using high concentration of Cu as the catalyst at 60°C [75]. Toor *et al.* finally achieved B-Si with a mean reflectivity of 3.1% using a two steps Cu-assisted etching method at 50°C for 5 min. They also demonstrated a solar cell with a conversion efficiency of 17.0% using Cu-assisted chemical etched nanoporous B-Si on pyramidal-textured Si substrate [77]. Figure 3.14 shows both top-view and cross-sectional SEM images of Cu-etched nanoporous Si surfaces at etch compositions ρ ($\rho=[HF]/([HF]+[H_2O_2])$) from 22% to 90% [77]. The ρ provides control of the nanostructure morphology; when ρ is above 60%, the density grade that reduces the reflectivity in the Au-assisted chemical etched or Ag-assisted chemical etched method is achieved. The relationship between the average reflectivity (R_{ave}) and ρ is shown in Figure 3.15; where the R_{ave} is reducing when $\rho < 80\%$, and the lowest R_{ave} of 3.1% is achieved at ρ~80% [77]. Moreover, this reflectivity range is comparable to that obtained using Au or Ag as the catalysts. Cao *et al.* successfully

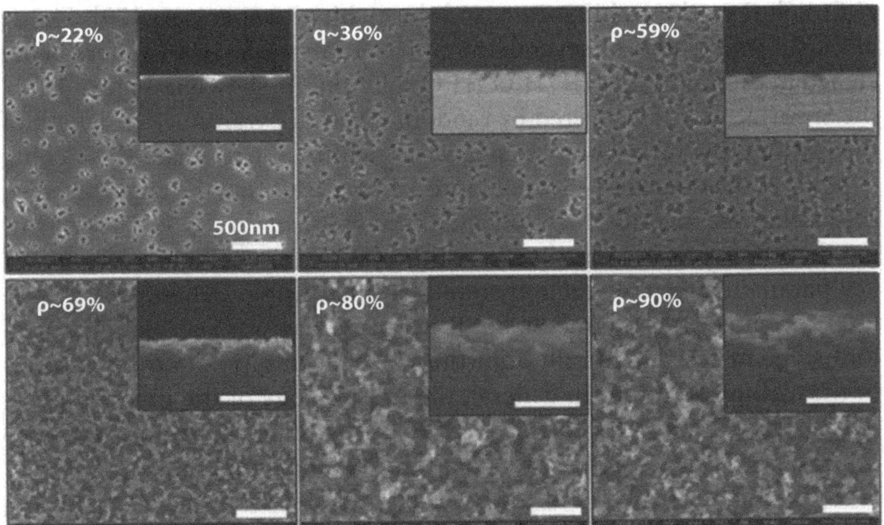

Figure 3.14 Top-view and cross-sectional (insets) SEM images of Cu-nanoparticle etched nanoporous Si as a function of $\rho = [HF]/([HF]+[H_2O_2])$. Scale bars are 500 nm. Adapted with permission [77]. Copyright 2014, John Wiley and Sons.

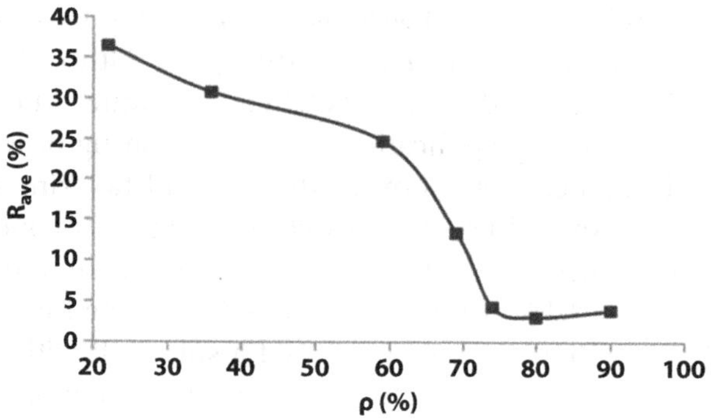

Figure 3.15 Solar-spectrum-weighted reflectance (R_{ave}) versus ρ = [HF]/([HF]+[H_2O_2]) for Cu-etched nanoporous B-Si. Adapted with permission [77]. Copyright 2014, John Wiley and Sons.

produced nanostructured porous layers or craters on Si surfaces *via* a one-step Cu-assisted chemical etching in a solution of HF/$Cu(NO_3)_2$/H_2O_2 at 40°C [78]. It is speculated that the micrometer-scaled craters that appear at high concentration of Cu^{2+} ions may be caused by the increased interaction between Cu^{2+} ions and the Si surface, resulting in enhanced lateral etching [78].

Unique inverted pyramid structured Si are fabricated by Wang *et al. via* adjusting the concentration of each chemical etchant in a HF/$Cu(NO_3)_2$/H_2O_2 solution, the etching time, and the etching temperature [79]. Figure 3.16a shows a photo of several 156 mm × 156 mm Si (100) wafers being etched in the $Cu(NO_3)_2$/HF/H_2O_2/H_2O solution. As shown in Figure 3.16b, the Si surface is fully covered by random inverted pyramids. The length of the inverted pyramids' bottom side varies within 2–6 μm, and the etching depth is in the range of 1–5 μm. Figure 3.16d reveals that the angle between the facet of the inverted pyramid and the Si (100) surface is 54.7°, confirming that the facets are terminated with Si (111) planes (Figure 3.16c). The key parameters of the reaction were identified using commercial, p-type (100)-oriented, 1–3 Ω·cm c-Si, but the process also was successfully tested for n-type material, naturally adjusting the concentration of the chemicals in the etching solution. In general, the morphology of the inverted pyramid arrays can be well controlled by adjusting the etching time, the concentration of $Cu(NO_3)_2$/HF/H_2O_2, and the etching temperature.

The reflectance spectra of the inverted pyramid arrays fabricated using different etching times for Wang *et al.* method are shown in Figure 3.17, illustrating the tunability of the Cu-assisted chemical etching technique. Notably, the inverted pyramid arrays can reduce the mean reflectivity of c-Si down to 5% over the wavelength range from 300 to 1,000 nm.

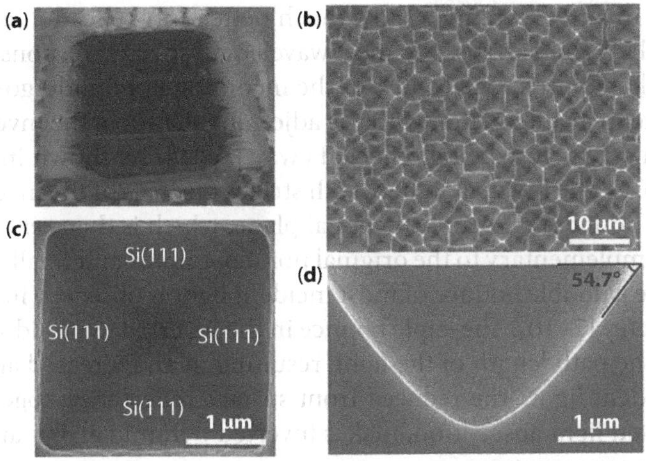

Figure 3.16 A photo of Cu nanoparticle- assisted etching process and SEM images. (a) 156 mm × 156 mm c-Si wafers being etched in a $Cu(NO_3)_2/HF/H_2O_2/H_2O$ etching bath, (b) SEM top-view image of the inverted pyramid arrays after 15 min processing, (c) SEM top-view image of an individual inverted pyramid processed for 15 min, (d) SEM cross-sectional view of an individual inverted pyramid processed for 15 min. Adapted with permission [79]. Copyright 2015, Nature Publishing Group.

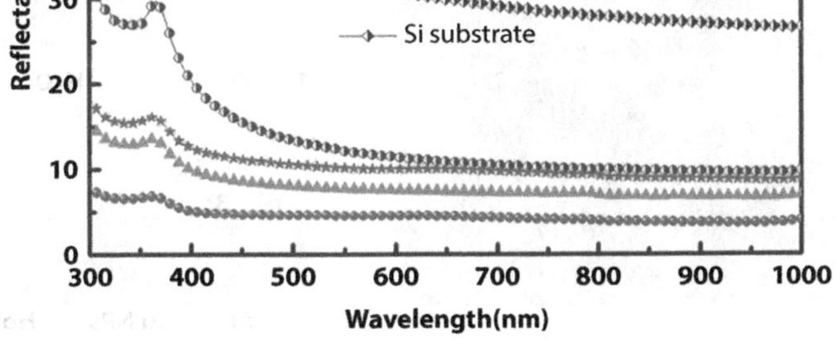

Figure 3.17 Reflectance spectra of pyramid arrays and the inverted pyramid arrays obtained using Cu nanoparticles assisted anisotropic etching for 5, 10, and 15 min. The inset shows schematics of the light path in the normal pyramidal structure and an individual inverted pyramid. Adapted with permission [79]. Copyright 2015, Nature Publishing Group.

Moreover, inverted pyramids show much better anti-reflection effect than the normal pyramid arrays. The short wavelength spectral response analysis indicates that approximately 40% of the incoming light undergoes a triple bounce due to its interaction with the adjoining planes of the inverted pyramid geometry before being reflected away [7–10]. As shown in the inset in Figure 3.17, the incident light, which strikes one face of the inverted pyramid, first bounces onto an orthogonal plane, which is then reflected onto the face complementary to the original point of impact and finally reflected away, while a double bounce of most incident light is observed in the pyramid geometry [7–10]. The triple bounce in the inverted pyramid structures increases the path length of the light, resulting in an increased absorption of the incident light. The reduced front surface reflectance, together with the increased light absorption, makes inverted pyramid arrays an efficient light trapping geometry.

In the technique of Wang *et al.*, the anisotropic deposition of Cu nanoparticles from the $Cu(NO_3)_2/HF/H_2O_2/H_2O$ mixture is the key step for the synthesis of the inverted pyramid arrays without a mask [79]. Figure 3.18 illustrates the mechanism of a one-step Cu-assisted chemical etching technique for fabricating inverted pyramid structures. It is an electrochemical reaction between Si and Cu^{2+}/Cu, similar to the metal-assisted chemical etching technique mentioned in Sections 3.3 and 3.4. The driving force of the reaction is the different electrochemical potentials between Si and Cu^{2+}/Cu [64, 68, 80, 81]. The reaction can be described as two half-cell reactions:

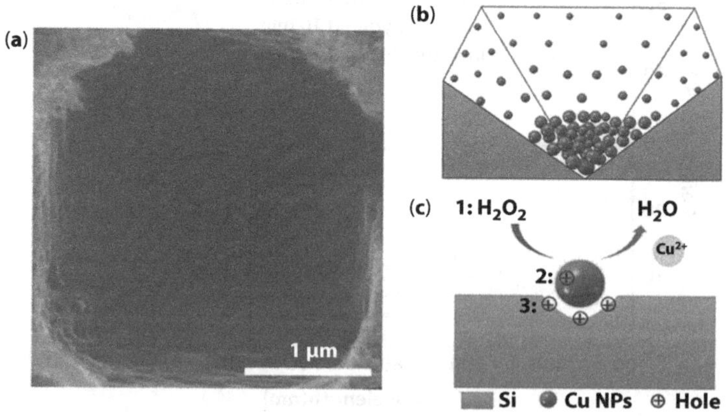

Figure 3.18 Scanning electron microscope image and schematics of Cu nanoparticle-assisted anisotropic etching. (a) SEM top-view image of an individual inverted pyramid after etching for 1 min, (b) Schematics of the anisotropic Cu nanoparticles' deposition, (c) Schematics of the H_2O_2 reduction process and the mechanisms for the holes injection. Adapted with permission [79]. Copyright 2015, Nature Publishing Group.

Cathode reaction:

$$Cu^{2+} + 2e^{-} \rightarrow Cu^{0}(s) \tag{3.7}$$

$$H_2O_2 + 2H^{+} \rightarrow 2H_2O + 2h^{+} \tag{3.8}$$

Anode reaction:

$$Si(s) + 2H_2O \rightarrow SiO_2 + 4H^{+} + 4e^{-} \tag{3.9}$$

$$SiO_2(s) + 6HF \rightarrow H_2SiF_6 + 2H_2O \tag{3.10}$$

In case of the reactions, there is no difference between Cu-assisted and other metal-assisted chemical etching. The Si/$Cu(NO_3)_2$/H_2O_2/HF system is composed of a corrosion-type redox couple; the cathodic reduction of Cu^{2+} ions with H_2O_2 as well as the anodic oxidation and dissolution of Si beneath the deposited Cu nanoparticles [64, 69, 80, 81]. The mechanism for the electrochemical depositing of metal nanoparticles is similar when taking single nanoparticle into account. However, Cu nanoparticles are anisotropically deposited on the Si surface when come to the group, which is different from the deposition of other metal nanoparticles. This is because the reduction potential is low for Cu^{2+} ions, limiting its ability to capture electrons, thus the Cu-assisted chemical etching process is carried out at 50°C in order to improve the ability of capturing electrons of Cu^{2+} ions. While the reduction potential of Ag^{+} ions, Au^{+} ions or other noble metal ions is much stronger than Cu^{2+} ions, exhibiting a strong ability of capturing electrons, therefore Ag-assisted chemical etching or Au-assisted chemical etching technique usually proceeds at room temperature and only takes several minutes. Moreover, the electronic properties of Si (100) and (111) planes are different in terms of their surface bond densities [82, 83]. Accounting for that, the number of electrons "available" on Si (100) plane is greater than that on Si (111) plane, so Cu^{2+} ions will preferentially capture electrons from the Si (100) planes making Cu nanoparticles formation more preferable, thereby resulting in the faster growth of Cu nanoparticles on Si (100) planes than that on Si (111) planes. Figure 3.18a shows an early form of one inverted pyramid structure [79], where more and larger Cu nanoparticles are observed on Si (100) planes and fewer and smaller Cu nanoparticles are observed on Si (111) planes. Figure 3.18b depicts a simulated diagram that shows the pattern of anisotropic Cu nanoparticles' deposition. As a result, Si under the Cu nanoparticles are preferentially oxidized and then etched by HF, leading to a faster etching along the Si

<100> directions. However, due to the strong ability of capturing electrons of Ag^+ ions, Au^+ ions, or other noble metal ions, there is no obvious difference in obtaining electrons from Si (100) and Si (111) planes, thus nanostructured B-Si are obtained.

Notably, H_2O_2 plays an important and special role in selective etching. Energetically, the electrochemical potential of H_2O_2 is much more positive than either the valence band of Si or the reduction electron potential of Cu^{2+}/Cu, and it is well-documented [11] that the reduction of H_2O_2 takes place on the surface of the metal, therefore, the reduction of H_2O_2 (labeled "1" in Figure 3.18c) in the Si/$Cu(NO_3)_2$/H_2O_2/HF system occurs on the surface of the Cu nanoparticles [84]. When moderate concentration of H_2O_2 is introduced, Cu nanoparticles are deposited anisotropically, which assists in the etching of inverted pyramids by HF; thus, the injected holes are consumed *via* two pathways. The holes are consumed by the Cu nanoparticles, ensuring kinetic equilibrium between Cu nanoparticles' deposition and dissolution (labeled "2" in Figure 3.18c). Some holes also diffuse through the Cu nanoparticles and are injected into the Si substrate (labeled "3" in Figure 3.18c) selectively accelerating the etching of c-Si beneath the Cu nanoparticles.

In the absence of H_2O_2, the reduction of Cu^{2+} ions is faster, allowing more Cu nanoparticles to be formed at the Si surface because of unlimited supply of Cu^{2+} ions. The Cu nanoparticles quickly grow larger and cover the majority of the Si wafer surface [85]. This situation induces overlapping regions of oxidation and, accordingly, more isotropic Si etching. A thin Cu film forms very quickly at 50°C and inhibits the etching of Si by acting as a barrier between the Si and the etchants. At high concentrations of H_2O_2, the oxidation of Cu nanoparticles is faster than the reduction of Cu^{2+} ions, resulting in the deposition of a few small random Cu nanoparticles on the Si surface; therefore, the original morphology of the Si does not substantially change, except for the formation of a few shallow pits. In summary, due to the limited ability of capturing electrons of Cu^{2+} ions, and the different rates of supplying electrons of Si (100) planes and Si (111) planes, combining the interaction between H_2O_2 and Cu nanoparticles, inverted pyramid structures are produced by a one-step Cu-assisted chemical etching.

These inverted pyramid structure obtained by Cu-assisted chemical etching technique not only shows good anti-reflection effect but also be facile passivated due to its micro-scale size compared to the upright pyramid structures obtained by conventional texturing method. Moreover, the bottom of the inverted pyramid structure is open and wide, making it very suitable for the conformal deposition of SiNx passivation film and the close contact with metal electrode. Thus, Si solar cells based on inverted pyramid structures obtained by Cu-assisted chemical etching process and upright pyramid structures obtained by conventional texturing method are fabricated and investigated systematically [86]. An 18.87% efficient inverted pyramid structured Si solar cell is achieved with open-circuit voltage of 637.7 mV, short-circuit current of 37.47 mA/cm^2 and fill factor of 78.84%, respectively [86]. (The performance of the Si solar cells fabricated by the company is relatively low compared to other companies.) In comparison with a similar processed conventional upright pyramid Si cells, the Cu-assisted chemical etched inverted pyramid Si cells collect 0.59 mA/cm^2 more short-circuit current and 0.47% more conversion efficiency on behalf of its superior light-trapping ability for absorbing more incident light and recessed structure characteristics for sufficient passivation and good ohm contact [86]. Furthermore, as compared to the Ag-assisted chemical etched and Au-assisted chemical etched technique, the cost of the Cu-assisted chemical etched method is much lower [86]. The results indicate that the inverted pyramid Si cells have enhanced light absorption and improved passivation and electrode contact [86]. As such, Cu-assisted chemical etching technique, compatible with current production line, is low cost and simple, which may trigger a new era of solar cells based on inverted pyramid structured Si [86].

This Cu-assisted chemical etched inverted pyramid structure is also used to fabricate hybrid organic-Si heterojunction solar cells. The influence of interface conformity was studied on the PEDOT:PSS/Si hybrid solar cells based on the conventional upright pyramid structure, the reprocessed rounded upright pyramid structure and the inverted pyramid structure [87]. As a result, a trend of increasing performance in the PEDOT:PSS/Si hybrid solar cells as a function of the interface conformity is revealed. Due to superior light trapping effect, the improved conformity of the Ag/organic/Si interface, together with the decreased serial resistance, the PEDOT:PSS/Si hybrid solar cells based on inverted pyramid structure shows the best performance, with a conversion efficiency of 10.04%, open-circuit voltage of 504 mV, short-circuit current of 30.59 mA/cm^2, fill factor of 65.12% and series resistance of 0.15 ohm·cm^2, respectively [87].

3.5.2 Copper-Assisted Chemical Etching of m-Si

For Si solar cells, 60% of the cost comes from the Si wafer, while m-Si costs 30%–50% less than c-Si, thus m-Si shares more than a half of the market [88, 89]. However, conventional slurry wire sawing (SWS) method to cut the Si ingots into wafers has many disadvantages, such as slow cut speed, waste generation, and high breakage of steel wire [90]. In order to further reduce the cost, diamond wire sawn (DWS) technique is introduced [91]. Although DWS method has many advantages such as a higher productivity, an easier recycling of the cooling liquid, and a lower wear of the wire [91–94] and DWS wafers show more than a factor of two smaller saw damage layer thicknesses compared to that of SWS wafers [90], the surface of the wafers produced by DWS is different from that produced by SWS, the conventional acid texturing method does not work well on DWS m-Si wafers and currently hinders further development of this innovative DWS technique [95, 96].

Inspired by the technique of Cu-assisted chemical etching of c-Si, this technique is introduced to texture the DWS m-Si wafer [97]. The DWS m-Si wafers textured by the Cu-assisted chemical etching technique and conventional acid texturing technique are investigated systematically and the solar cells based on these two kinds of wafers are studied. Because m-Si has grains with different directions, the obtained structures are not perfect inverted pyramids, except for the <100> grain. However, the 22.4% mean reflectance of the Cu-assisted chemical etched DWS m-Si wafer is 8.4% lower than that for the conventional acid textured DWS m-Si wafer (30.8%) over the wavelength range from 300 to 1,100 nm, moreover, due to the low reflectance and high light absorption because of the inverted pyramid like structures, the solar cell based on the Cu-assisted chemical etched DWS m-Si wafer show a much higher conversion efficiency of approximately 19.33%, as well as a higher short-circuit current of 37.42 mA/cm^2, compared to solar cell based on the conventional acid textured DWS m-Si wafer, which shows that of 18.83% and 36.69 mA/cm^2, respectively [97].

3.6 Conclusion

Metal-assisted chemical etching has been proposed as advanced texturing and has been introduced in detail from the definition, mechanisms, and phenomena during the etching and the influences of etchants. It has been developed from two-step method to one-step method, which greatly reduces the energy cost and process complexity. What is more, one-step metal-assisted chemical etching method has further decreased the energy cost by using Cu, Ag instead of Au as the metal catalyst. There is no big difference in the features of the obtained structures by Au-etched or Ag-etched. Cu, as the cheapest metal among these metal catalysts, has attracted significant attention. At first, it was difficult to obtain B-Si by Cu-assisted chemical etching. Afterward, by adjusting the composition of the etchants and the etching temperature, B-Si was finally achieved by Cu-assisted chemical etching. By further adjusting the etching conditions, inverted pyramid structures was firstly produced by Cu-assisted chemical etching. In a word, B-Si with different morphologies can be simply obtained by metal-assisted (such as Au, Ag, and Cu) chemical etching method, but by using Cu-assisted chemical etching method, not only B-Si but also inverted pyramid structures can be achieved.

It is well accepted that metal-assisted chemical etching is an electrochemical reaction, and the mechanisms of metal-assisted chemical etching can be tentatively attributed to holes injection during the etching process. Holes are selectively injected at the interface of the metal/Si. On the one hand, the injected holes will be consumed by the Si underneath the metal nanoparticles; on the other hand, the injected holes will diffuse to other regions if the rate of holes consuming is slower than the rate of holes injection. Then, the Si substrate are oxidized and etched by the HF, resulting in the formation of Si nanostructures. However, there are still some open questions: 1). Where H_2 is from? 2). How Si is dissolved? 3). Why different metal catalysts show different catalytic activities? Up to now, the proper explanation of some currently poorly understands the phenomena, which relies on fully understanding of holes injection process. Even though the mechanisms of metal-assisted chemical etching are not fully confirmed, the etching phenomena of Si are controllable and reproducible. The mechanism will be investigated in more detail in the future along with the development of technology.

Minimization the reflection losses can effectively improve the performance of Si solar cells; thus, the study of texturing process is attracting attention of researchers. Conventional texturing can reduce the reflectivity down to ~10% for c-Si and ~20% for m-Si in the wavelength range 300 to 1,000 nm. Moreover, the reflectivity can be further reduced by advanced texturing, namely, metal-assisted chemical etching. What is more, metal-assisted chemical etching method is simple and low cost, and the resulted structures show good light-trapping effects, which make it very promising in the application of Si solar cells. The application of metal-assisted chemical etching method for m-Si solar cells has been mass industrialized, and the application of this method in the mass production of c-Si solar cells is under the way.

References

1. Singh, P.K., Kumar, R., Lal, M., Singh, S.N., Das, B.K., *Sol. Energy Mater. Sol. Cells*, 70, 103, 2001.
2. Vazsonyi, E., De Clercq, K., Einhaus, R., Van Kerschaver, E., Said, K., Poortmans, J., Szlufcik, J., Nijs, J., *Sol. Energy Mater. Sol. Cells*, 57, 179, 1999.

3. Kim, H., Park, S., Kang, B., Kim, S., Tark, S.J., Kim, D., Dahiwale, S.S., *Appl. Surf. Sci.*, 284, 133, 2013.
4. Wang, F., Zhang, X., Wang, L., Jiang, Y., Wei, C., Zhao, Y., *J. Power Sources*, 293, 698, 2015.
5. Seidel, H., *J. Electrochem. Soc.*, 137, 3626, 1990.
6. Nishimoto, Y., Ishihara, T., Namba, K., *J. Electrochem. Soc.*, 146, 457, 1999.
7. Smith, A.W. and Rohatgi, A., *Sol. Energy Mater. Sol. Cells*, 29, 37, 1993.
8. Baker-Finch, S.C. and McIntosh, K.R., *Prog. Photovolt. Res. Appl.*, 19, 406, 2011.
9. Baker-Finch, S.C. and McIntosh, K.R., *Prog. Photovolt. Res. Appl.*, 20, 51, 2012.
10. Campbell, P. and Green, M.A., *J. Appl. Phys.*, 62, 243, 1987.
11. Huang, Z., Geyer, N., Werner, P., de Boor, J., Gösele, U., *Adv. Mater.*, 23, 285, 2011.
12. Otto, M., Algasinger, M., Branz, H., Gesemann, B., Gimpel, T., Füchsel, K., Käsebier, T., Kontermann, S., Koynov, S., Li, X., Naumann, V., Oh, J., Sprafke, A.N., Ziegler, J., Zilk, M., Wehrspohn, R.B., *Adv. Opt. Mater.*, 3, 147, 2015.
13. Liu, X., Coxon, P.R., Peters, M., Hoex, B., Cole, J.M., Fray, D.J., *Energy Env. Sci.*, 7, 3223, 2014.
14. Chourou, M.L., Fukami, K., Sakka, T., Virtanen, S., Ogata, Y.H., *Electrochimica Acta*, 55, 903, 2010.
15. Um, H.-D., Kim, N., Lee, K., Hwang, I., Hoon Seo, J., Yu, Y.J., Duane, P., Wober, M., Seo, K., *Sci. Rep.*, 5, 11277, 2015.
16. Zhang, M.-L., Peng, K.-Q., Fan, X., Jie, J.-S., Zhang, R.-Q., Lee, S.-T., Wong, N.-B., *J. Phys. Chem. C*, 112, 4444, 2008.
17. Hadjersi, T., *Appl. Surf. Sci.*, 253, 4156, 2007.
18. Huang, Z., Zhang, X., Reiche, M., Liu, L., Lee, W., Shimizu, T., Senz, S., Gösele, U., *Nano Lett.*, 8, 3046, 2008.
19. Chattopadhyay, S., Li, X., Bohn, P.W., *J. Appl. Phys.*, 91, 6134, 2002.
20. Peng, K., Yan, Y., Gao, S., Zhu, J., *Adv. Funct. Mater.*, 13, 127, 2003.
21. Peng, K. and Zhu, J., *J. Electroanal. Chem.*, 558, 35, 2003.
22. Peng, K. and Zhu, J., *Electrochimica Acta*, 49, 2563, 2004.
23. Peng, K.Q., Hu, J.J., Yan, Y.J., Wu, Y., Fang, H., Xu, Y., Lee, S.T., Zhu, J., *Adv. Funct. Mater.*, 16, 387, 2006.
24. Koynov, S., Brandt, M.S., Stutzmann, M., *Appl. Phys. Lett.*, 88, 203107, 2006.
25. Sai, H., Fujii, H., Arafune, K., Ohshita, Y., Yamaguchi, M., Kanamori, Y., Yugami, H., *Appl. Phys. Lett.*, 88, 201116, 2006.
26. Sai, H., Fujii, H., Arafune, K., Ohshita, Y., Kanamori, Y., Yugami, H., Yamaguchi, M., *Jpn. J. Appl. Phys.*, 46, 3333, 2007.
27. Nishioka, K., Horita, S., Ohdaira, K., Matsumura, H., *Sol. Energy Mater. Sol. Cells*, 92, 919, 2008.
28. Nishioka, K., Sueto, T., Saito, N., *Appl. Surf. Sci.*, 255, 9504, 2009.
29. Lin, C.-H., Dimitrov, D.Z., Du, C.-H., Lan, C.-W., *Phys. Status Solidi C*, 7, 2778, 2010.

30. Hsu, C.-H., Wu, J.-R., Lu, Y.-T., Flood, D.J., Barron, A.R., Chen, L.-C., *Mater. Sci. Semicond. Process.*, 25, 2, 2014.
31. Sai, H., Kanamori, Y., Arafune, K., Ohshita, Y., Yamaguchi, M., *Prog. Photovolt. Res. Appl.*, 15, 415, 2007.
32. Archer, R.J., *J. Phys. Chem. Solids*, 14, 104, 1960.
33. Kooij, E.S., Butter, K., Kelly, J.J., *Electrochem. Solid-State Lett.*, 2, 178, 1999.
34. Dimova-Malinovska, D., Sendova-Vassileva, M., Tzenov, N., Kamenova, M., *Thin Solid Films*, 297, 9, 1997.
35. Li, X. and Bohn, P.W., *Appl. Phys. Lett.*, 77, 2572, 2000.
36. Huang, Z., Fang, H., Zhu, J., *Adv. Mater.*, 19, 744, 2007.
37. Fang, H., Wu, Y., Zhao, J., Zhu, J., *Nanotechnology*, 17, 3768, 2006.
38. Yae, S., Kawamoto, Y., Tanaka, H., Fukumuro, N., Matsuda, H., *Electrochem. Commun.*, 5, 632, 2003.
39. Tsujino, K. and Matsumura, M., *Adv. Mater.*, 17, 1045, 2005.
40. Tsujino, K., Matsumura, M., Nishimoto, Y., *Sol. Energy Mater. Sol. Cells*, 90, 100, 2006.
41. Peng, K., Wu, Y., Fang, H., Zhong, X., Xu, Y., Zhu, J., *Angew. Chem. Int. Ed.*, 44, 2737, 2005.
42. Marnix, V., de Goeij, J.J., Kolar, Z.I., Frens, G., Lusse, P.J., Zuiddam, M.R., van der Drift, E., *J. Electrochem. Soc.*, 148, C28, 2001.
43. Yuan, H.-C., Yost, V.E., Page, M.R., Stradins, P., Meier, D.L., Branz, H.M., *Appl. Phys. Lett.*, 95, 123501, 2009.
44. Tsujino, K. and Matsumura, M., *Electrochimica Acta*, 53, 28, 2007.
45. Toor, F., Branz, H.M., Page, M.R., Jones, K.M., Yuan, H.-C., *Appl. Phys. Lett.*, 99, 103501, 2011.
46. Zheng, H., Han, M., Zheng, P., Zheng, L., Qin, H., Deng, L., *Mater. Lett.*, 118, 146, 2014.
47. Peng, K.Q., Huang, Z.P., Zhu, J., *Adv. Mater.*, 16, 73, 2004.
48. Qiu, T., Wu, X.L., Siu, G.G., Chu, P.K., *J. Electron. Mater.*, 35, 1879, 2006.
49. Srivastava, S.K., Kumar, D., Schmitt, S.W., Sood, K.N., Christiansen, S.H., Singh, P.K., *Nanotechnology*, 25, 175601, 2014.
50. de Boor, J., Geyer, N., Wittemann, J.V., Gösele, U., Schmidt, V., *Nanotechnology*, 21, 095302, 2010.
51. Peng, K.-Q., Yan, Y.-J., Gao, S.-P., Zhu, J., and others, *Adv. Mater.*, 14, 11e64, 2002.
52. Choi, W.K., Liew, T.H., Dawood, M.K., Smith, H.I., Thompson, C.V., Hong, M.H., *Nano Lett.*, 8, 3799, 2008.
53. Harada, Y., Li, X., Bohn, P.W., Nuzzo, R.G., *J. Am. Chem. Soc.*, 123, 8709, 2001.
54. Chartier, C., Bastide, S., Lévy-Clément, C., *Electrochimica Acta*, 53, 5509, 2008.
55. Chen, C.-Y., Wu, C.-S., Chou, C.-J., Yen, T.-J., *Adv. Mater.*, 20, 3811, 2008.
56. Lee, C.-L., Tsujino, K., Kanda, Y., Ikeda, S., Matsumura, M., *J. Mater. Chem.*, 18, 1015, 2008.

57. Peng, K., Fang, H., Hu, J., Wu, Y., Zhu, J., Yan, Y., Lee, S., *Chem. - Eur. J.*, 12, 7942, 2006.
58. Ogata, Y., Kobayashi, K., Motoyama, M., *Curr. Opin. Solid State Mater. Sci.*, 10, 163, 2006.
59. Xia, X.H., Ashruf, C.M.A., French, P.J., Kelly, J.J., *Chem. Mater.*, 12, 1671, 2000.
60. Asoh, H., Sakamoto, S., Ono, S., *J. Colloid Interface Sci.*, 316, 547, 2007.
61. Ono, S., Oide, A., Asoh, H., *Electrochimica Acta*, 52, 2898, 2007.
62. Peng, K., Lu, A., Zhang, R., Lee, S.-T., *Adv. Funct. Mater.*, 18, 3026, 2008.
63. Unagami, T., *J. Electrochem. Soc.*, 127, 476, 1980.
64. Norga, G.J., Platero, M., Black, K.A., Reddy, A.J., Michel, J., Kimerling, L.C., *J. Electrochem. Soc.*, 144, 2801, 1997.
65. Huang, Z.P., Geyer, N., Liu, L.F., Li, M.Y., Zhong, P., *Nanotechnology*, 21, 465301, 2010.
66. Cruz, S., Hönig-d'Orville, A., Müller, J., *J. Electrochem. Soc.*, 152, C418, 2005.
67. Lu, Y.-T. and Barron, A.R., *J. Mater. Chem. A*, 2, 12043, 2014.
68. Morinaga, H., Suyama, M., Ohmi, T., *J. Electrochem. Soc.*, 141, 2834, 1994.
69. Mitsugi, N. and Nagai, K., *J. Electrochem. Soc.*, 151, G302, 2004.
70. Tsujino, K. and Matsumura, M., *Electrochem. Solid-State Lett.*, 8, C193, 2005.
71. Branz, H.M., Yost, V.E., Ward, S., Jones, K.M., To, B., Stradins, P., *Appl. Phys. Lett.*, 94, 231121, 2009.
72. Milazzo, R.G., D'Arrigo, G., Spinella, C., Grimaldi, M.G., Rimini, E., *J. Electrochem. Soc.*, 159, D521, 2012.
73. Oh, J., Yuan, H.-C., Branz, H.M., *Nat. Nanotechnol.*, 7, 743, 2012.
74. Liu, Y., Lai, T., Li, H., Wang, Y., Mei, Z., Liang, H., Li, Z., Zhang, F., Wang, W., Kuznetsov, A.Y., Du, X., *Small*, 8, 1392, 2012.
75. Yu, D., Wang, S., Li, L., Wang, W., Wu, S., Wu, X., Zhang, F., *Micronanoelectron. Technol.*, 249, 2014.
76. Lee, J.-P., Choi, S., Park, S., *Langmuir*, 27, 809, 2011.
77. Toor, F., Oh, J., Branz, H.M., *Prog. Photovolt. Res. Appl.*, 23, 1375, 2015.
78. Cao, Y., Zhou, Y., Liu, F., Zhou, Y., Zhang, Y., Liu, Y., Guo, Y., *ECS J. Solid State Sci. Technol.*, 4, 331, 2015.
79. Wang, Y., Yang, L., Liu, Y., Mei, Z., Chen, W., Li, J., Liang, H., Kuznetsov, A., Xiaolong, D., *Sci. Rep.*, 5, 10843, 2015.
80. Teerlinck, I., Mertens, P.W., Schmidt, H.F., Meuris, M., Heyns, M.M., *J. Electrochem. Soc.*, 143, 3323, 1996.
81. Kim, J.S., Morita, H., Joo, J.D., Ohmi, T., *J. Electrochem. Soc.*, 144, 3275, 1997.
82. Hesketh, P.J., *J. Electrochem. Soc.*, 140, 1080, 1993.
83. Yang, L., Liu, Y., Wang, Y., Li, X., Chen, W., Hua, Y., Zhang, Q., Fu, J., Liang, H., Mei, Z., Du, X., *RSC Adv.*, 4, 24458, 2014.
84. Selvaraju, T. and Ramaraj, R., *J. Appl. Electrochem.*, 39, 321, 2009.
85. Chyan, O.M.R., *J. Electrochem. Soc.*, 143, 92, 1996.
86. Yang, L., Liu, Y., Wang, Y., Chen, W., Chen, Q., Wu, J., Kuznetsov, A., Du, X., *Sol. Energy Mater. Sol. Cells*, 166, 121, 2017.

87. Yang, L., Liu, Y., Chen, W., Wang, Y., Liang, H., Mei, Z., Kuznetsov, A., Du, X., *ACS Appl. Mater. Interfaces*, 8, 26, 2016.
88. Möller, H.J., Funke, C., Rinio, M., Scholz, S., *Thin Solid Films*, 487, 179, 2005.
89. Kumagai, A., *Sol. Energy Mater. Sol. Cells*, 133, 216, 2015.
90. Watanabe, N., Kondo, Y., Ide, D., Matsuki, T., Takato, H., Sakata, I., *Prog. Photovolt. Res. Appl.*, 18, 485, 2010.
91. Meinel, B., Koschwitz, T., Acker, J., *Energy Procedia*, 27, 330, 2012.
92. Meinel, B., Koschwitz, T., Blocks, C., Acker, J., *Mater. Sci. Semicond. Process.*, 26, 93, 2014.
93. Yang, C., Wu, H., Melkote, S., Danyluk, S., *Adv. Eng. Mater.*, 15, 358, 2013.
94. Meng, H. and Zhou, L., *Silicon*, 6, 129, 2014.
95. Lippold, M., Buchholz, F., Gondek, C., Honeit, F., Wefringhaus, E., Kroke, E., *Sol. Energy Mater. Sol. Cells*, 127, 104, 2014.
96. Fashina, A.A., Adama, K.K., Oyewole, O.K., Anye, V.C., Asare, J., Zebaze Kana, M.G., Soboyejo, W.O., *J. Renew. Sustain. Energy*, 7, 063119, 2015.
97. Wu, J., Liu, Y., Chen, Q., Chen, W., Yang, L., Wang, Y., He, M., Du, X., *Sol. Energy*, 171, 675, 2018.

4

Wet Chemical Cleaning for Industrial Application

Florian Buchholz[1]*, Eckard Wefringhaus[1] and Martin Plettig[2]

[1]International Solar Energy Research Center Konstanz (ISC Konstanz) e.V., Constance, Germany

[2]RENA Technologies GmbH, Gütenbach, Germany

Abstract

The wet chemical cleaning of wafer surfaces is required after several process steps in current state-of-the-art silicon solar cell production technology. Apart from the cleaning efficiency, process stability, cost, and throughput considerations have to be met. The goal of this chapter is to introduce and discuss the most common cleaning techniques within the current and potential future solar cell process sequences.

An important part when discussion wafer cleaning technology is contamination: what kind of contamination is present at which step and what are the consequences for the single process steps and the overall production chain. Hence, we introduce the concept of "contamination management"—referring to the knowledge of the main contamination sources and its variations as basis for a deeper understanding for process variation and also process optimization. As literature on the topic is relatively rare, we provide experimental examples that may be used as starting point for further studies. Also, the provided techniques and data may be used to introduce contamination monitoring procedures in the industrial production environment. In a last section, we will discuss the relevance of cost for the ongoing evolution in wet chemical production technology.

***Keywords*:** Solar cell processing, industrial scale cleaning, surface contamination

**Corresponding author*: florian.buchholz@isc-konstanz.de

Monika Freunek Müller (ed.) Photovoltaic Manufacturing: Etching, Texturing, and Cleaning, (115–160) © 2021 Scrivener Publishing LLC

4.1 Introduction

Silicon solar cell and module manufacturing has drastically changed over the recent years and so has the price prices per Watt peak power (the common pricing unit). Figure 4.1 depicts this decline in sales prices with increased shipments and the respective years. The solar industry has seen the rise and decline of the German solar module manufacturing sector triggered by the feed-in tariff system and the huge rise of cell and module manufacturing in China. Yet, with some bumps, the trend has remained. With increasing shipment of modules, the prices per unit have decreased, making solar electricity one of the cheapest forms of electricity in many parts of the world with electricity costs (levelized costs of electricity, LCOE) of below 0.10 USD/KWh (utility scale) [1]—a decrease to a third of the price in 2010.

As being an offshoot of semiconductor manufacturing, the solar cell industry has emancipated itself over this drastic increase in production capacity. The process technology has developed with the focus on increasing throughput and minimizing process times and costs. So, processes were altered and, in particular in wet chemistry, vastly simplified and the high standards of cleanliness, known from the semiconductor industry, were strongly reduced. Yet, the solar cell is still a device made from silicon, and with efficiencies and complexities more and more increasing, the

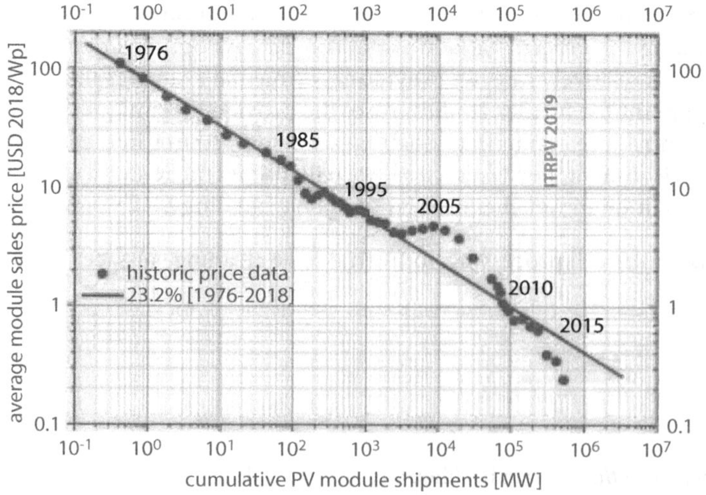

Figure 4.1 Average sales price reduction plotted against the cumulated PV module shipments [2].

sensitivity to common impurities, such as transition metal surface contamination, is also expected to increase. Accordingly, wet chemical cleaning of the silicon surface has become an integral part in the manufacturing.

As wafer cleaning is closely related to the etching of wafers, this chapter is dedicated to both. In the first section, the status of production technology in solar cell manufacturing is outlined. Subsequently, common practice in wet chemical processing is summarized—tools, etching of silicon, cleaning of wafer surfaces, rinsing and drying, and how they are integrated into the overall process flow. Despite of their crucial impact on solar cell performance, the general knowledge on contaminants is mostly rudimentary, not many comprehensive publications exist. From experience, in factories, the topic only tends to pop up when no other reasons for process failures can be discovered. Thus, in the main section of this chapter, the concept of "contamination management" is introduced. If cleaning processes are to be studied, contamination needs to be quantified. Accordingly, first the question of how to measure contamination is addressed. Then, the main sources of metal contamination are identified: the as-cut wafer and alkaline etching steps. Data on (process induced) variations is provided to give an idea on the stability and variability of the system. Two metal contamination species are identified to serve as "contaminants of interest" for process control and a method for monitoring them cost effectively is introduced. Apart from metal contamination, also organic residues on the wafer can impact the performance of a solar cell production line, which will be discussed next. Finally, in this section, a cost-effective method for cleaning of wafer surfaces (based on HF/ozone solutions) is introduced and it is shown how it experimentally was optimized for best performance. Finally, we will give an overview on the cost structure of the most common solar cell process sequence, the impact of the introduction of ozone-based cleaning will be discussed and further cost saving potentials will be assessed.

4.2 Status of Production Technology in Solar Cell Manufacturing

There are three trends to be observed in current industrial scale solar cell manufacturing that contribute to the constantly falling sales prices. First, economy of scale [3]—the largest solar cell and module production companies (i.e., the members of the Silicon Module Super League, SMSL) today have production capacities in the range of ten gigawatt, which translates into almost 100 running production lines—leaving the rest of the world struggling behind [4]. Second, related to the first is an ever increasing throughput of the production equipment to reduce the prices per wafer. Nowadays, machines are required to have a throughput of 7,000 wafers per hour and more [2]. Third, in order to further reduce the costs of the final produced electricity by the solar systems (LCOE), the efficiency of the produced solar cells needs to be continuously increased. The last aspect has led to a remarkable increase in innovation in a formerly highly conservative branch of industry. The last 20 years were dominated by the same type of solar cell technology, namely the aluminum back surface solar cell (Al-BSF). For example, SMSL super league member Hanwha Q-cells claims an efficiency increase of $0.6\%_{absolute}$ per year for the last ten years [5]. Factors that allowed for such an increase are the availability of high quality silicon, improvements in the performance of screen printing pastes, continuous process optimization, and lately, a shift in technology. The Al-BSF technology, mainly on cheap multicrystalline silicon substrates, which dominated the marked for years, is increasingly replaced by passivated emitter and rear cell (PERC) technology [6]. One reason for the success of the PERC technology is that it was essentially introduced as an upgrade to existing lines—the so-called non-disruptive approach. First, one existing production line could be upgraded so that the technology could prove its stability before it was introduced in large scale. Now, as PERC is reaching its limits, the successor technology is yet to be determined. Disruptive approaches are discussed, however, from the history of the industry, the authors would rather expect a cell technology that again can be introduced by smaller alteration of the current state-of-the-art production, instead of disruptive approaches.

One key factor for the success of a technology in mass production is the stability of each of the single process steps. The smaller the number of cells with low performance, the higher the yield of the line. Also, process optimization is simplified if processes have a reproducible outcome. Another important factor is the involved costs, i.e., the capital expenditure (CAPEX) and the operating expenditures (OPEX). Both are related to throughput—the higher the throughput of the machines, the lower the cost per wafer.

As Chinese companies have become dominant in the production of solar cells, another "soft" factor has to be considered when discussion production technology: trends. Once a technology is regarded suitable for some reason or another by one Chinese manufacturer, a growing demand throughout the industry can occur. This increasing demand in turn can help the technology to mature, and eventually it may change the global production reality.

4.3 Wet Chemical Process Technology

4.3.1 Tools

Etching and cleaning steps in solar cell production are commonly implemented into the same tools. Depending on the treatment time, mostly governed by the etching bath, inline- or batch-type tools are used. In order to keep constant treatment conditions, fresh chemicals are added and used chemicals are dumped in pre-defined intervals ("feed-and-bleed processing"), which may be after a certain wafer throughput number and/or after a fixed time (idle or in production). Excess treatment solution is commonly extracted via an overflow in the bath or in the corresponding tank.

In batch-type tools, wafers are stacked in carriers—commonly one hundred pieces per carrier. For wet processing, typically plastic carriers are used. If oxidative chemistry is used, fluoridated polymers are required, in other cases less expensive polymers can be utilized. Within the machine currently up to six carriers are combined to batches and transported from bath to bath and immersed into the baths by a robot. The throughput is governed by the number of carriers (wafers) per batch and the longest treatment time. To increase throughput, when longer treatment times are required, redundant baths are added. Currently, throughput with four carriers (400 wafers) per batch is in the range of 7,000 to 8,000 wafers per hour.

There are different types of treatment baths (the letters refer to Figure 4.2):

- Simple dipping bath (a). Carriers are dipped into the bath. Nitrogen or clean dry air (CDA) bubbling may be optionally used for improved mixing.
- Recirculated bath (b). For tempering and/or gas injection (via a static or membrane mixer), the bath is circulated.
- Overflow rinse (c). The inlet of the fresh water is at the bottom of the rinse, the flow of water is constant during processing. Nitrogen or CDA bubbling optional.
- Dump rinsing bath (d). To reduce the rinsing time, the water is quickly dumped via a large valve at the bottom, after the carrier has entered the bath. This can be combined with a spraying unit refilling the bath, while the carrier is still in it [c.f. (e)].
- Spray rinsing bath (e). Rinsing water is introduced via spray nozzles.

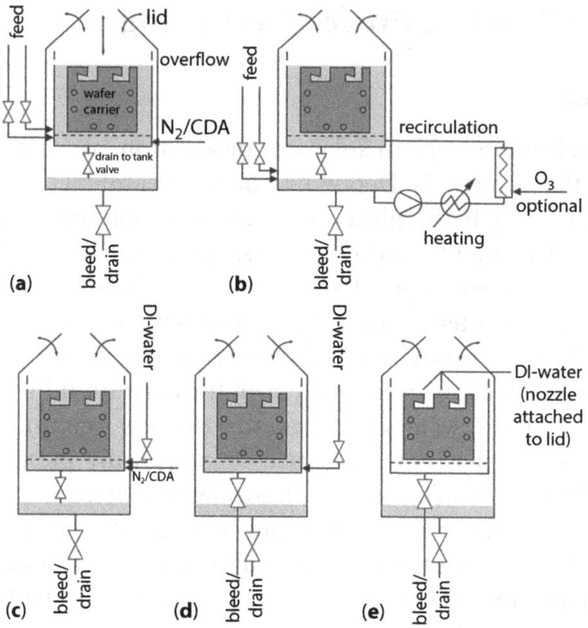

Figure 4.2 Common process and rinsing baths for batch processing. (a) simple dipping bath, (b) recirculated bath, (c) overflow rinsing bath, (d) dump rinsing bath, (e) spray rinsing bath.

Bath types (a) and (b) are mainly used for etching and cleaning. Simple dipping baths are commonly used for HF or HF/HCl dips. Recirculated baths are required, when the solution needs to be heated (e.g., for alkaline texturing or saw damage etching) or gas (typically ozone) needs to be injected. Only two feed lines are sketched. Depending on the bath make-up (i.e., the number of mixed chemicals), more can be required.

Thorough rinsing with deionized (DI) water is required to remove carry-over of process chemicals to avoid contamination and stains on wafer surfaces. Possible bath setups are sketched in (c) to (e). Overflow rinses with optional bubbling function are the most common types. Dump rinsing may be necessary to speed up processing times, particularly when sticky chemicals are used. The wafers are dipped into the filled bath, and then, the water is dumped quickly and the rinse is re-filled. Alternatively, the dumping is programmed to happen after the treatment time, in order to get to low conductance values again. Spray rinsing is an alternative to save water. Cascading of rinsing baths is another option for water consumption minimization and reduction of processing times. The quality of the rinsing process is commonly determined by electrical conductance measurement.

Drying of wafers is achieved by hot water and hot air driers. A typical batch etching tool is shown in Figure 4.3.

In inline-type machines wafers are transported via rotating conveyor rolls from bath to bath. Automation equipment places the wafers on lanes (typically five) to reach a throughput of up to 6,000 wafers per hour. An image of an inline tool can be found in Figure 4.4. For single side etching, the wafers "float" on the etching solution. To achieve this, the etching bath level is adjusted by varying the pump power of the circulation pump in such a way that it is just high enough to contact the lower side and the edge of the wafer, while the wafer is still resting on the rolls (e.g., on O-rings). In this way, a meniscus of etching solution is formed between the wafers surface and the etching bath, allowing to etch only the lower side of the wafer.

The up-facing side can be covered by a protecting water cap to avoid etching from the gas phase. The water cap is obtained by spraying a defined amount of water, which should be small enough to not mix with etching solution to avoid dilution. A sketch for a single side bath can be found labeled with the letter (a) in Figure 4.5. For both sided treatment, wafers can be led through spraying baths as is shown in (b) or immersion baths. Spraying baths are mostly used for cleaning or rinsing, while immersion baths are most common for both sided silicon etching (such as acidic texturing).

Immersion bath treatment can be realized by transporting the wafers through the bath by altering the transportation direction from the horizontal axis (c). More elegantly, as the horizontal transportation axis can be kept throughout the machine, this can also be achieved by using a similar setup as in (a), but using a stronger pump. The bath level can then be lifted above the level of the conveyor rolls to create a "mound" of bath solution, which the wafers then enter. In both cases, to prevent the wafers from

Figure 4.3 Batch tool, on left side inlet for carriers. Robot arm with four carriers can be seen. ©RENA Technologies GmbH.

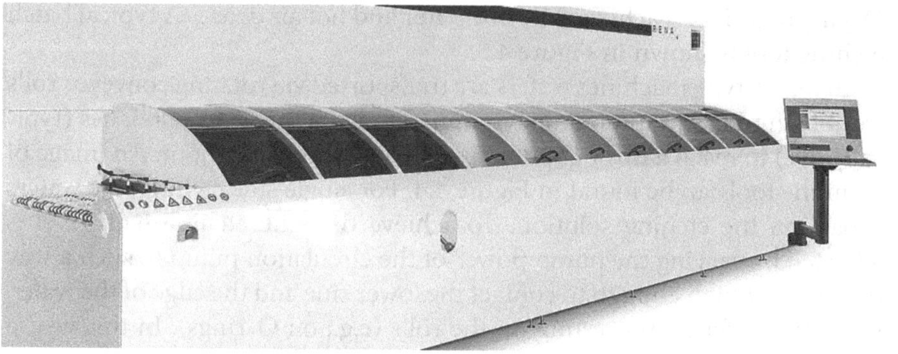

Figure 4.4 Inline tool, on left side automation for wafer placement on the lanes needs to be added. ©RENA Technologies GmbH.

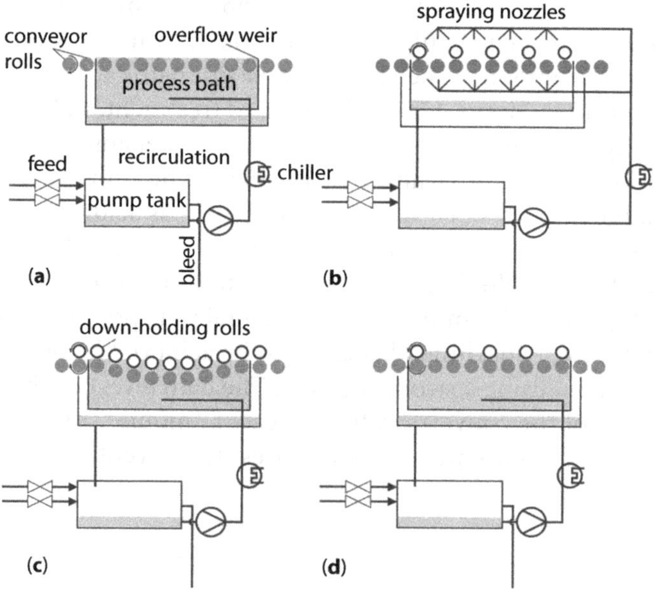

Figure 4.5 Common process and rinsing for inline processing. Wafers are transported via conveyor rolls. (a) single side processing bath, (b) spraying bath, (c) immersion bath, (d) both side treatment bath with horizontal wafer transportation, using overflow weirs.

floating, down-holding rolls are used. The down-holding force is created by the weight of loosely to the roll attached elements.

Acidic etching of silicon is a vigorous exothermic reaction. To keep the process controlled and avoid excessive heating of the etching solution, heat exchangers are used. For cleaning and rinsing baths no tempering is required and thus the heat exchanger left out. Principally, it is also possible to heat the etching solution, if necessary.

4.3.2 Etching

The most important etching step of the silicon surfaces is the removal of the saw damage from the wire sawing process of the as-received solar wafer. This is often done in combination with a surface texture etch producing a "rough" surface of some kind to improve the light trapping properties of the final device.

Depending on the kind of substrate and the purpose of the silicon etching steps different etching systems are used. Acidic isotropic etching (based on HNO_3/HF mixtures) has been widely used to texture multicrystalline (mc) wafers [7, 8]. However, the acidic etching is highly sensitive to the surface structure before etching. With the introduction of diamond wire sawing of wafers replacing the use of the loose abrasive (SiC) slurry technique, which entails significant ecological and cost benefits [9], the acidic etching of the wafer surfaces are no longer yielding the high quality surface structures. One of the problems that entails the diamond wire sawing technology is the parallel grooves on the wafer surface that are still visible after etching [10]. Thus, there have been various alternative approaches for mc wafer texturing, using high sulfuric acid concentrations [11] or by metal catalized etching [12, 13]. The latter uses metal (in most publications silver) precipitates to locally increase the etching rate. To avoid a negative effect of those particles thorough cleaning is required.

Excellent surface texturing results are obtained when immersing monocrystalline (mono) wafers as obtained by the Czochralski process (Cz) [of (100)-crystal orientation] into diluted alkaline solutions containing an organic additive. Randomly distributed pyramids of roughly the same size (3–10 µm in height) are formed. The first substance for widespread use as texturing additive was iso-propanol [14, 15]. A whole range of chemical classes have has patented ever since: dioles, other alcohols (almost of every kind), glycoles, alcoxyglycoles, sodiumcarbonate, but also all sorts of other surfactants and generally all sorts of organic molecules with -OH groups. The main motivation of introducing new additives has been that the vapor pressure of iso-propanol is high; thus, significant amounts are lost through the gas phase. Accordingly, only relatively low temperatures can be used leading to long treatment times. Today, highest surface homogeneity, lowest reflection and process reliability can be obtained by commercially available additive mixtures from various vendors.

Without the organic additive and preferably in higher concentrated alkaline solution "flat" polished surfaces are obtained. Highest etch rates well above 1 µm/min are found for temperatures exceeding 80°C in the range between 15 and 20%wt. KOH. The etch rate is independent for various alkali metal cations (Na^+, K^+, Li^+) [16].

As the etch rate of alkaline solutions is substantially lower as compared to acidic solutions, both alkaline texturing and polishing require heating of the solutions and a batch process. In order to match the throughput targets, multiple baths of the same type in parallel can be used.

The second common etching step is performed after the emitter diffusion process. Wafers are treated single sidedly to remove the emitter from the rear side of the device. Depending on the properties of the etching solutions this process step may be used to polish the rear surface of the substrate to improve the passivation quality. For single side polishing, more concentrated solutions are used. With comparatively high etching rates, acidic etching is most commonly performed in inline etching tools [17, 18], allowing short treatment times below one minute for the etching. In the following list the most common etching processes are summarized.

- Saw damage etching and polishing of Cz wafers: KOH or NaOH, concentrated solution (≈20%$_{mass}$) heated to 80°C and above [16, 19], mostly in batch-type equipment.
- Saw damage etching and texturing of Cz wafers (random pyramids): KOH or NaOH, diluted solution (1%–5%), organic additive, heated to 80°C and above, in batch-type equipment [20–22].
- Saw damage etching and texturing of mc wafers (slurry sawed): HNO_3 (≈40%), HF (≈10%) in CH_3COOH or H_2O, cooled etching solution, inline-type equipment [23].
- Texturing of mc wafers (diamond wire sawed): metal catalyzed etching based on silver nano particles from solution, etching in HF/H_2O_2, various ways of process [12, 13].
- Single side polishing of mc or Cz wafers in HNO_3 (≈50%), HF (≈10%), H_2SO_4 (≈30%), cooled etching solution, inline-type process.

Apart from the silicon etching, etching of silicate glasses, such as phosphorous silicate glass (PSG) or boron silicate glass (BSG) are common in most process sequences. These are performed in diluted hydrofluoric acid. Furthermore, porous silicon that is formed during acidic etching needs to be removed—the common method for this is a quick dip or rinse in diluted KOH solution.

4.3.3 Cleaning

Silicon etching may have two effects on the silicon surface in terms of cleanliness. On the one hand, it removes previously present contamination or, on the other hand, it may add to the surface contamination content. Most commonly, three different types of contaminants are differentiated: inorganic molecular scale contamination, ions, uncharged compounds or atoms; organic contamination (molecules, thin films or layers); and particles (silicon-based particles or dust from the surrounding). Of the inorganic contamination species, transition metals are of highest interest. Once trapped (physisorbed or chemisorbed) at the surface of the silicon wafer, they can easily penetrate the silicon bulk, some even at relatively low temperatures [24]. Inside the bulk or at the interface, they pose the threat of being minority charge carrier recombination sources [25–27]. Defect-induced recombination of charge carriers prevents the carriers from reaching the contacts; accordingly, the efficiency of the device is reduced. Inorganic contamination sources can be, for example, impure chemicals, contact of the wafer with metal surfaces, and dust from corrosion of metal parts of machines. Major source for organic contaminants are sawing residues, such as glues, but also fingerprints. Insufficient removal of these residues is known in particular to interfere with alkaline etching [28–30], but also other process steps.

Acidic etching, such as acidic texturing, cleans the wafer surface very effectively due to the high solubility of metal contaminants in low pH and high oxidizing potential solutions [31, 32]. However, when removing porous silicon residues caused by gaseous phase etching upon the exit of the etching bath, which is commonly done in diluted alkaline solution of KOH or NaOH, the wafer surface gets contaminated by metallic residues. The reason for this is that the common "cheap" bases with similar properties cannot be obtained in similarly high purity as acids (as the pure substance is a solid, distillation is not possible) and the high pH in addition to that strongly limits the solubility of most metal contamination species [33].

For a long time, wafers were only dipped into diluted acids (HCl and HF mixtures) for cleaning purposes, even after alkaline etching steps [34]. Yet, with solar cell efficiencies increasing, more efficient cleaning technology was required. In order to account for the removal of more precious elements such as copper (which seemed to be hardly of relevance with the Al-BSF technology [35]) and to remove organic residues, oxidative cleaning baths, followed by oxide removal steps, have become widely used. Drawing from semiconductor manufacturing technology, the RCA cleaning sequence would be one obvious choice. However, the sheer number of process bath and the reaction conditions required in this sequence

[SC1: hydrogen peroxide/ammonium hydroxide mixture at 80°C, rinsing, (diluted HF), rinsing; SC2: hydrogen peroxide/hydrochloric acid mixture at 80°C, rinsing, (diluted HF), rinsing and drying] are hardly feasible for solar cell mass production. Apart from the number of process steps, the high consumption of chemicals is a major disadvantage. While the SC1 mainly targets organic contamination and particles, the SC2 step is dedicated for the removal of metal impurities [36]. Simplified versions of this cleaning sequence, however, are widely used in the solar industry e.g.:

- pSC1 (pseudo SC1), diluted KOH and H_2O_2 mixture (≈1% and 1%) at elevated temperature, without HF step for the cleaning of as-cut wafers or for the preparation of wafers for texturing, or followed by a HCl/HF step (≈2%/2%) for cleaning after alkaline texturing [37].
- SC2 only (commonly with subsequent HF dip), used after saw damage etching (alkaline polishing) or for metal removal when necessary [38].

However, hydrogen peroxide mixtures have the drawback of being quite expensive. First of all, hydrogen peroxide at sufficient purity is quite expensive; second, it is consumed relatively heavily, as the stability of the molecule is low at higher temperatures and in the presence of metal impurities that can serve as catalyst for the decomposition [39]. Particularly in alkaline solutions, this can become a severe problem, as the heated alkaline solutions start etching (polishing) the surface if not enough hydrogen peroxide is present. The relative costs for chemicals of a alkaline texturing and cleaning wetbench with pre- and post-clean based on hydrogen peroxide are plotted Figure 4.6.

With the goal of reducing the running costs, attempts have been made to replace hydrogen peroxide by ozone as an oxidizing agent. Ozone is commonly produced from clean oxygen supply using ozone generators as part of the etching and cleaning tools and dissolved in the cleaning baths via static mixing units (a process that has also been first employed in the semiconductor industry). For more information on ozone-based cleaning in the semiconductor industry and a first attempt using it for solar cell manufacturing, refer to [40–48]. As the majority of the semiconductor literature is on HF/O_3, we have spent some time to optimize the cleaning system containing diluted HF and ozone. An example for such an optimization experiment will be given at the end of the next section.

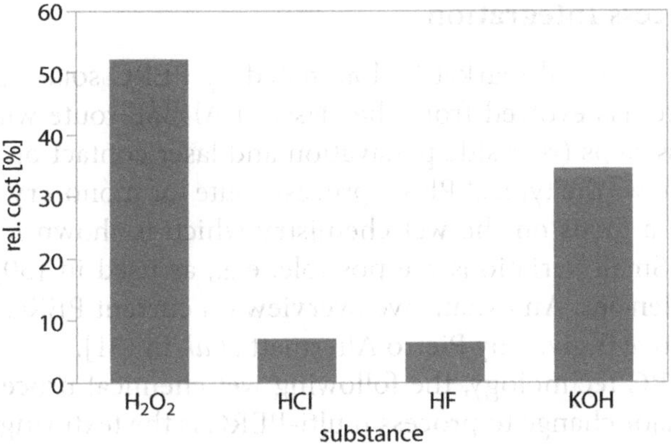

Figure 4.6 Relative costs for chemicals in a state-of-the-art wetbench, with a batch size of 400 wafers and throughput of 6,000 wafers per hour. PSC1 cleaning is used before and after the texturing step. Cost for texturing additive excluded.

4.3.4 Rinsing and Drying

Usually, each etching and cleaning step is followed by rinsing in DI water to remove the remains of the respective chemicals. There may be more than one rinse used in sequence to shorten process times and to reduce water consumption by using the rinsing baths in a cascade. Before the wafers are dried at the end of the machine, the surfaces are rendered hydrophobic by immersing the wafers into diluted HF or HF/HCl mixtures. While the effect of drying has been studied in depth in the IC industry, for solar cell manufacturing to our knowledge, no literature on best drying practice exists.

Generally, water stain formation needs to be avoided and process time needs to be reduced as much as possible. Most rapid drying can be achieved by blow-drying of hyrophobic wafer surfaces (which does not only allow fast drying but also avoids stain formation). Thus, in inline-type tools, blown-drying with hot air knifes is the common drying method. A similar effect is obtained by lifting the wafer carriers in batch-type tools through a horizontal hot air stream. In batch-type equipment prior to air drying, a hot water bath may be beneficial to pre-heat the wafers and carriers, yet negative effects of the hot water drier have been observed in the field, in particular for highly surface sensitive process sequences (for hetero-junction solar cells).

4.3.5 Process Integration

The current solar cell market is dominated by PERC solar cells [2]. The process route has evolved from the classical Al-BSF route with only two extra process steps (rear side passivation and laser contact opening) [49]. Figure 4.7 shows the typical PERC process route for mono-crystalline substrates, with a focus on the wet chemistry which is shown in increasing complexity. Small variations are possible, e.g., as used in [50], to obtain highest efficiencies. An exhaustive overview on current PERC technology and perspective is given by Pietro Altermatt *et al.* in [51].

In Cz-PERC technology, the following wet chemical process steps are used (the major change to process multi-PERC is the texturing process):

- A batch-type wetbench for pre-cleaning, alkaline texturing, and post-cleaning.
- An inline-type wetbench for the single side etching (polishing and rear side emitter removal), porous silicon etching, subsequent cleaning, and PSG removal.

With very deep saw damage, it has been beneficial to remove the saw damage first and then run the alkaline texturing step [37, 53].

Other industrial solar cells, such as bifacial n-type solar cells (for example, ISC Konstanz' BiSoN solar cells [54, 55]) can be produced by splitting

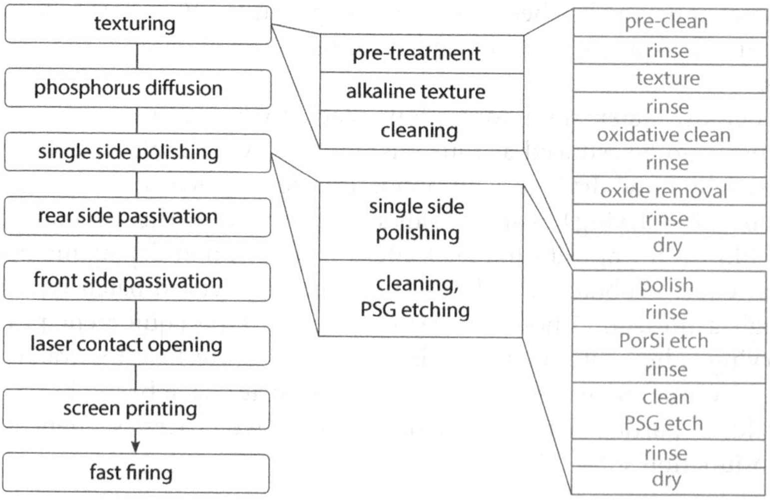

Figure 4.7 Typical state-of-the-art PERC process route with the wet chemical process steps in increasing degrees of complexity [37, 52].

and moreover pulling apart the saw damage etching and texturing within the process sequence. In this case, the texturing of the wafers can be moved further toward the back end of the line (back end of line, BEOL). First, the saw damage of the surface is removed. Then, the surfaces are cleaned and diffused. After single side etch barrier deposition, the wafers are textured on the non-covered side. In order to reduce the process complexity, SiN_x (or AlO_x/SiN_x) layers may be used as barrier and kept to the end of the process to serve as rear side passivation layers. In this case, the etching and cleaning steps need to be adjusted so that only minimal damage is done to the barrier layer.

Another example where two etching steps follow each other is acidic inline etching. Here, a rinse in diluted KOH is necessary to etch the film of porous silicon that forms by gas-phase etching upon leaving the acidic etching bath. A general scheme on how etching and cleaning steps are integrated into solar cell processes of all kinds is shown in Figure 4.8. Apart from etching and cleaning, pre-cleaning step may be used to prepare the surface for the etching, e.g., by removing potential organic residues or interfering silicate glass or other layers that may be present.

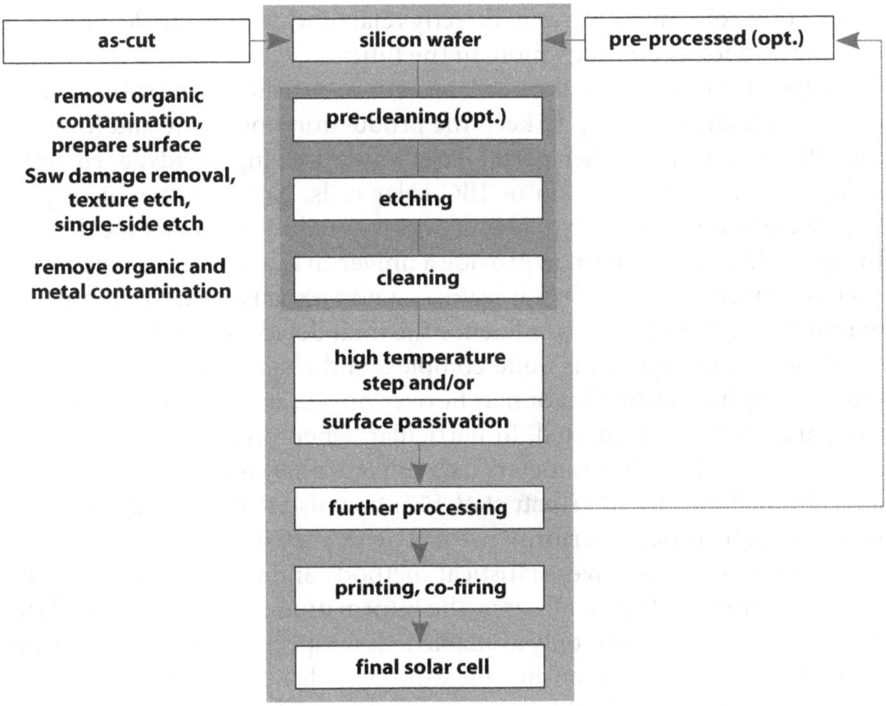

Figure 4.8 General scheme of how wet processes are integrated into the solar cell process.

4.4 Contamination Management

As stated by Altermatt *et al.* in [51], cleanliness is gaining importance for the yield of solar cell production. So far, improvement in solar cell performance has been achieved mainly based on fine-tuning of existing processes. In particular, in the back-end of the solar cell process, screen printing has shown incredible performance gains in the last years. At the same time, diffusion profiles and passivation layers have been optimized for homogeneity and performance. Wet chemical process steps and the used machines have not changed much in the last few years. The replacement of IPA as texturing additive by commercially available additives, which helped to substantially improve process stability and homogeneity can be considered and exception. Not much has changed in respect to the fabrication surrounding. Commonly, some sort of "clean-room" technology is used. ISO-7 has become the unofficial standard. However, particle counts are rarely checked and clean room discipline is often low. In particular, during move in of machines, or when heavy works are required, the tools may be prone to all sorts of contamination. In particular, when cleaning baths get contaminated, cleaning them can become very challenging. In general, we believe that contamination and, directly related with it, wafer cleaning will be a topic to focus on much more in the future.

Deeper knowledge on the contamination measurement, sources, and countermeasures can help to keep the production line stable and improve the efficiency level further particularly when moving to advanced technology, such hetero-junction or IBC solar cells. Not only may it help to improve processes but also to identify process failure and machine defects. In Figure 4.9, we have tried to provide a universal scheme centering the wet chemical treatment of wafers. It features a non-exhaustive list of input and output parameters that may influence the final device properties.

Obviously, the system is quite complex, and changes at one end of the process may impact another or may be overlooked, as the same process control parameters are impacted, in particular, when only I-V-measurement (indicated as solar cell parameters data in Figure 4.9) is used. Nevertheless, it has been shown by Altermatt *et al.* [57] that also from this data, conclusions on single process performance can be extracted. Yet, this approach is quite complex and involves statistical methods and device modeling. The major downturn is that in all cases, the information on the process and the potential process issues is only available in retrospect, which is why process control based on off-/and on-line measurement devices cannot be replaced.

In this section, the highlighted parameters added to the process flow scheme in Figure 4.9 are discussed. First of all, it is important to measure

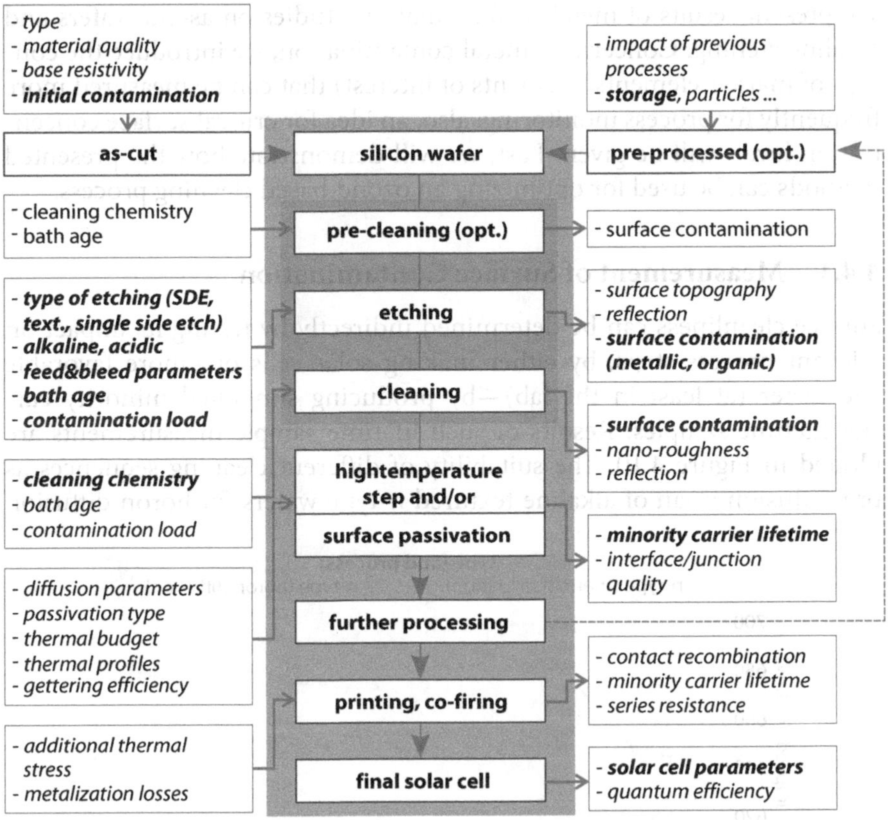

Figure 4.9 Summary of the relevant process steps centring on wafer cleaning. It includes an (incomplete list) of in- and output process parameters. Parameters that will be discussed in this chapter have been highlighted in bold italic typeface. Adapted from [56].

the cleanliness of a wafer in order to get an understanding of the origin of the contamination and how the contamination level is affected by etching cleaning steps. For this purpose, two methods for the determination of wafer cleanliness are introduced. In the following sections, we will provide measurement results and findings of extensive wafer surface analysis studies.[1] Of course most results, particular on incoming material properties and investigations in running process lines can only be considered as snapshots. Using the presented methods and results, the reader may come up with more advanced studies on the topics and/or individual solutions.

[1]This section is partly taken from the first author's dissertation with the title: Metal Surface Contamination in C-Si Solar Cell Processing, completed and defended 2016, TU Bergakademie Freiberg, Freiberg, Germany; which can be accessed here: https://nbn-resolving.org/urn:nbn:de:bsz:105-qucosa-203660.

We present results of metal surface analysis studies on as-cut wafers and alkaline etching. Concerning metal contamination, we introduce the concept of marker elements (elements of interest) that can be measured more frequently for process monitoring; also, an idea for critical surface concentration levels will be given. Last, we will demonstrate how the presented methods can be used for optimizing an ozone-based cleaning process.

4.4.1 Measurement of Surface Contamination

Surface cleanliness can be determined indirectly by testing its impact on relevant process steps, by either making solar cells or—more favorable and faster (at least in the lab)—by producing simplified minority carrier lifetime samples. Results of such lifetime sample measurements are plotted in Figure 4.10. The suitability of different cleaning sequences as pre-diffusion clean of alkaline textured n-type wafers for boron diffusion

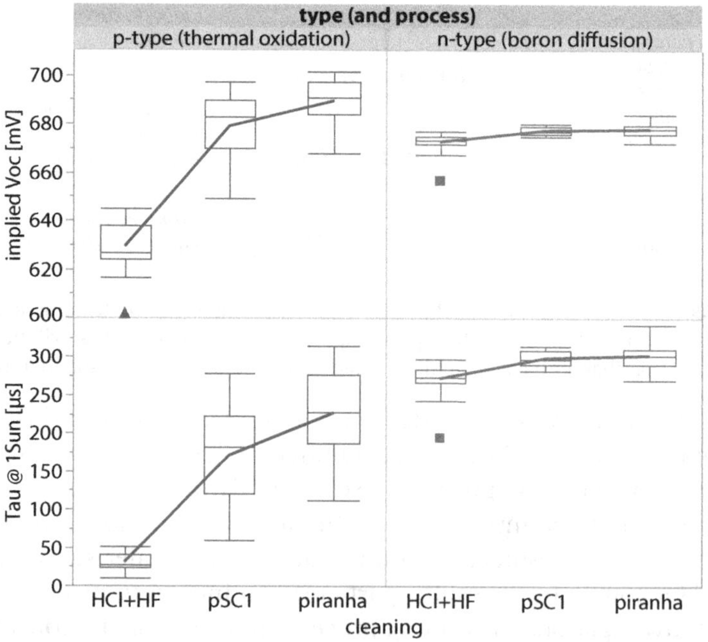

Figure 4.10 Impact of cleaning efficiency measured via thermally oxidized p-type and boron diffused n-type wafers. The group "HCl+HF" was cleaned using a 2% HF bath for 5 minutes, rinsed, and subsequently dipped in 2% HF for 2 minutes. The "pSC1" group was immersed into KOH/H_2O_2 (1/1%) at 60°C for 10 minutes and subsequently dipped in HF. The last group ("piranha") saw the pSC1 step and the HCl+HF step. Then, the wafers were immersed in sulfuric acid/hydrogen peroxide mixture 4:1 (SPM) at 80°C for 10 minutes and into 2% HF for 2 minutes. Last, all groups were dried in nitrogen [56].

(right-hand side) was tested. In addition similarly treated p-type wafers were thermally oxidized for comparison. Minority carrier lifetime samples were produced by subjecting the wafers to the respective high temperature step (thermal oxidation or boron diffusion) and passivating both sides of the wafer by PECVD SiN_x. After fast firing (a common step to "activate" SiN_x passivation), the minority carrier lifetime (tau) and the implied open circuit voltage (implied Voc) was measured by quasi-steady state photoconductance (QSSPC) measurement. The latter is a useful measure as it takes the base doping into account and can be related to final device Voc values (losses on lifetime samples will show at similar extend in final solar cell devices [58, 59]). From the graphs, it can be deduced that the tested process may not be sensitive enough toward contamination in order to discriminate between the tested cleaning sequences. For this, we use the thermal oxidation on p-type wafers, a process step that is highly sensitive toward surface contamination. The high temperatures drive the contaminants from the surface into the bulk increasing the bulk recombination or will enrich them at the surface increasing the surface recombination. The two graphs indicate the relevance and higher sensitivity of the thermal oxidation process, which makes it very useful for process optimization.

This method delivers the net influence of damage by the surface contamination but does not discriminate between the single contaminating species. As well known from semiconductor research, contamination levels may be in the range of down to 1e+9 cm^{-2}. Spatially resolved measurement is hardly possible at that low surface concentration values. A net measurement of the surface contamination content in the semiconductor industry is usually performed by vapor phase decomposition droplet collection method (VPD-DC), in which the wafer is placed inside a closed chamber on a cooled chuck. The chamber is flooded with HF fumes that condensate on the wafer surface and dissolve the native oxide. The condensate is then collected by scanning the wafer surface with a drop of HF/H_2O_2 containing solution. This works perfectly on flat wafer surfaces. However, as surfaces in solar cell manufacturing are rarely flat, we developed a method that also works on textured surfaces—the Droplet Sandwich Etch (DSE) method [31].

For this, the surface is etched from two sister wafers (wafers from the same batch) at the same time in order to extract the elements contained in the surface and surface near regions of the silicon wafers. A schematic description of the sampling method can be found in Figure 4.11. A droplet of 700 to 1,000 µl of diluted HF in H_2O_2 (HF/HNO_3 mixtures may also be used [60]) is placed on flat or textured surfaces, respectively (1). Highly pure (ultrapure) chemicals are required to enable a sufficiently low detection

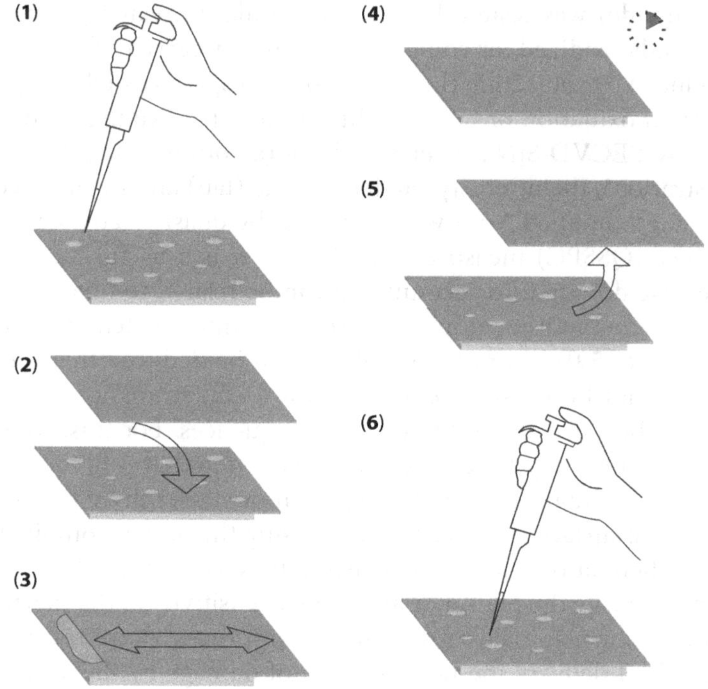

Figure 4.11 Droplet sandwich etch (DSE) sampling method for the determination of metallic and organic contaminaton [31].

limit[2]. A sister wafer is placed on top (2), the wafers are rubbed against each other (3) and the solution is left to soak for ten minutes (4). Then, the top wafer is removed (5), the solution is collected by a pipette (6) and the metal content is determined by ICP-MS analysis for multi-element mass spectrum analysis of a predefined selection of elements. ICP-MS analysis is capable of measuring many elements at the same time, so in principle the array of elements can be very big. Yet, due to the need of calibration spectrum data, the price scales with the number of tested elements. Recovery rates are generally very high for most elements as confirmed by Meyer *et al.* [61], with the exception of Cu—at least when spiked, due to the effect of high solubility and near room temperature diffusion of Cu in Si.

The DSE method has been further used to also determine the total carbon content of the surface as indicator for the organic surface contamination. The organic layer that may be present on the surface is extracted using the sandwich etch and the resulting solution is then analyzed using a TOC

---

[2]1ppb in sampling solution roughly corresponds to 1e+10atoms cm^{-2}.

instrument, which oxidizes the carbon by ozone to CO_2 and then measures the CO_2 by nondispersive infrared spectroscopy (NDIR) [60, 62].

4.4.2 As-Cut Wafer

It has been noted that slurry sawed as-cut wafers are the main source for metal contamination introduction in the solar cell process [31, 34]. In the following, the impact of the sawing technique on the initial surface contamination is discussed in more detail. In this section, the loose abrasive slurry sawing technique is compared to the fixed diamond particle abrasive wire sawing technique, which has become the standard sawing technique [9]. Furthermore, the homogeneity of surface metal contamination within an ingot is checked in order to get an idea of input variable variation, when treating as-cut wafers.

For this purpose, one column of each, n-type and p-type, 156×156 mm^2 pseudosquare Cz-silicon wafers was cut in half. One half was sliced into wafers by diamond wire sawing, the other half was sliced conventionally by the slurry sawing technique. The p-type wafers were used for the surface contamination studies. The homogeneity of the surface contamination across the column was investigated by analyzing wafers from both ends of the column. The distribution and the respective IDs (all wafers were laser coded for further processing) of the analyzed wafers can be found in Figure 4.12.

The surface concentration of an array of elements of the two different sawing techniques is plotted in Figure 4.13. Mo is not detected in either of the sawing techniques, which is good, as according to one of the first and most extensive studies about metal impurities by Davis *et al.* from 1980 [25], Mo was found to be the most effective solar cell "efficiency killer".

Of the other transition elements the highest concentrations were measured for Fe and Cu (up to 2e+13 and 2e+14cm^{-2}). Cu concentrations are highest on slurry sawed wafers, which was to be expected, as the steel wires

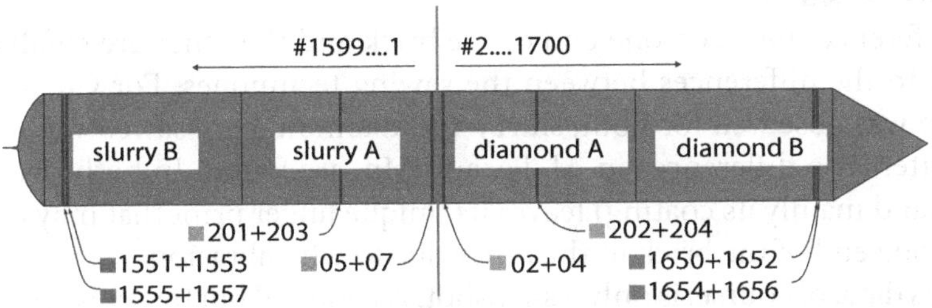

Figure 4.12 Column prepared for the analysis of the impact of the sawing process. The colors and wafer ID numbers correspond to surface contamination measurement results from Figure 4.13. About 700 to 800 wafers were cut from each half of the brick. From [56].

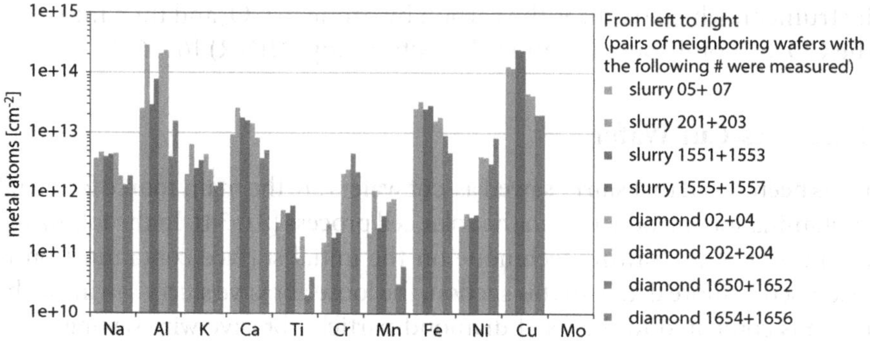

Figure 4.13 Surface contamination of as-cut wafers. The numbers of the sampled wafers refer to Figure 4.12 and are to be read from left to right [56].

used for the slurry sawing process are usually Cu coated. Interestingly, the Cu concentration of the diamond wires is still relatively high (>2e+13cm^{-2}). The base material of both types of wires appears to be steel, as Fe shows up in comparably high concentrations on both kinds of wafers. Ni and Cr are found to a similar extend as Fe on the diamond wire sawed wafers. On the slurry wire sawed wafers, the Ni and Cr concentration is about one order of magnitude lower. From the high Ni concentrations on the diamond wires, it may be concluded that the wires were copper-nickel coated. Mn and Ti follow in terms of magnitude and can just like Cr be a part of the coating alloy or part of the steel making up the core of the wire.

Of the lighter elements, highest concentrations of Al were measured. However, the variation between values is relatively high (a phenomenon observed throughout this work). Similar amounts for both sawing techniques are found of Na, K, and Ca. Over the origin of these high concentrations of alkali and earth alkali elements and Al can only be speculated. As the concentrations look similar, the post sawing treatment would be most likely the cause instead of sawing process related contamination sources, such as the slurry.

Differences between one end of the brick and the other are small compared to the differences between the sawing techniques. For Cu, a factor of two was observed for both, slurry and diamond wire sawed wafers. For the latter, also differences in Al, Fe, and Mn, are found. In both cases, the wire (and mainly its coating) leaves its unique finger print that may change from one end of the brick to the other due to wire abrasion.

This data, of course, is only a snapshot. As pointed out by Schweckendiek *et al.* [34], there is quite a variety of surface contamination to be expected when analyzing wafers from different manufacturers, as there may be different sawing processes and types of wires on the market. Apart from this, the sawing conditions are expected to have a significant impact on the outcome. The largest differences in the presented investigation—up to one order of magnitude—were measured for Ni, Ti, and Cr. The main contaminants (of the transition metal group) with the highest concentrations, however, are the same: Fe and Cu.

4.4.3 Alkaline Etching

This section aims to link the data from the previous section with the dynamics concerning metal contamination in the initial alkaline etching step. Most high efficiency solar cell processes start with alkaline etching in order to remove the saw damage of the as-cut wafer. In the following, the impact of the texturing etch on surface contamination is investigated with a special interest on the dynamics of surface contamination over the lifetime of alkaline etching baths. Last, the difference between alkaline etching at different points of the solar cell process sequence is addressed.

Wafers from the previous section were etched in NaOH (20%, Honeywell Puranal) at 80°C for the amount of time that was required to remove 9 to 10 μm per side. The batch size was 50 wafers, volume and bath size were in a similar relationship as in large scale production equipment. The solution was heated and circulated for homogeneous etching. As the concentration was not adjusted over time, the reduction of the etch rate with consumption of NaOH was compensated by increasing the etching time, due to practical reasons. This is in contrast to industrial applications, where the etch rate is kept constant by adding fresh chemicals (feed) and removing some of the old etching bath (bleed). After etching, the wafers were rinsed in an overflow rinse with DI water until a resistivity of 2 MΩ·cm was reached. The experiment was repeated using different wafers from the previous section to check reproducibility.

As shown in Figure 4.14, the surface contamination on the wafer surface after etching changes drastically over the first several carriers. The same array of elements as in the as-cut wafer analysis section before was investigated. The data fits are just guides for the eye. For the lighter elements, no trend was found. For the other elements, two different types of species can be extracted according to the main origin of the contaminants: the etching bath or the as-cut wafers. They can be discriminated by the fact that the surface concentration of the element is higher before or after etching of the first batches.

A great deal of Fe, Ni, and Mo was found on the wafer surface after etching stem from the etching bath. The bath is near its saturation point, alternatively, a (chemical or physical) reaction with the wafer surface can be

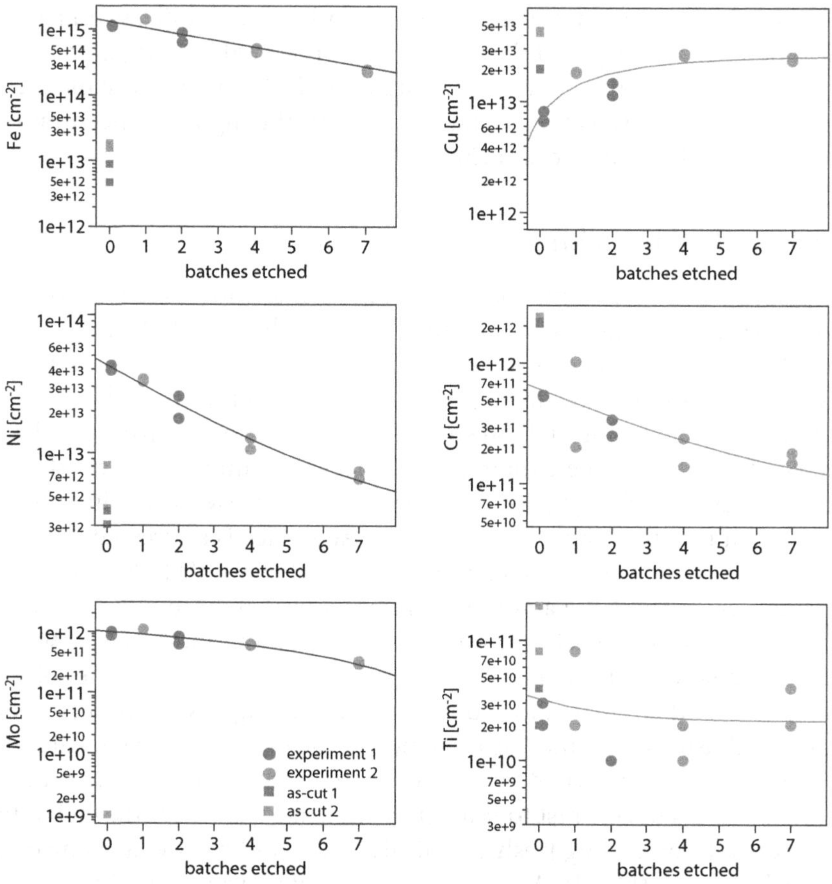

Figure 4.14 Surface contamination before (squares) and after (circles) etching. Left-hand side: elements stemming from the etching bath; right-hand side: elements stemming from the as-cut wafers. The diamond wire sawed p-type wafers from the previous section were used in two experiments (in the first experiment, wafers with small digit IDs, and in the second, with high digit IDs were used, cf. Figure 4.12). The initial surface contamination (before SDE) can be found in Figure 4.13. Taken from [56].

considered as reason for the outplating. Either way, the wafers will "clean" the bath from these species.

Assuming that no amount of the respective species A is found on the wafers when entering the etching bath, the "reaction equation" will look like the following:

$$A_{solution} \rightarrow A_{surface} \tag{4.1}$$

Depending on the deposition rate on the wafer surface the bath will be cleaned more or less quickly. In general, one can assume that some kind of equilibrium between the species in solution and the wafer surface will form (with an equilibrium constant k), which is dependent on the concentration in solution, solubility, and the diffusivity of the species A, but also on its (chemical or physical) adhesion to the wafer surface. Thus applies:

$$A_{solution} \rightleftharpoons A_{surface} \tag{4.2}$$

If the same species is present on the wafer upon entering the etching bath $A_{surface,i}$, this initial amount $A_{surface,i}$ can be added to the solution—as several micrometers of the surface are etched under gas bubble formation one can assume that the initial surface completely dissolves. Subsequently, during etching the above mentioned equilibrium will form. Thus applies:

$$A_{surface,i} \rightarrow A_{solution} \rightleftharpoons A_{surface} \tag{4.3}$$

A scheme that visualizes initial surface contamination, bath concentration, and surface contamination load after etching can be found in Figure 4.15. Depending on the kinetics of the adhesion to the wafer surface, diffusion, circulation of the bath, and other factors, the equilibrium will be either on side of the wafer (k > 1) or the bath (k < 1). The side of the

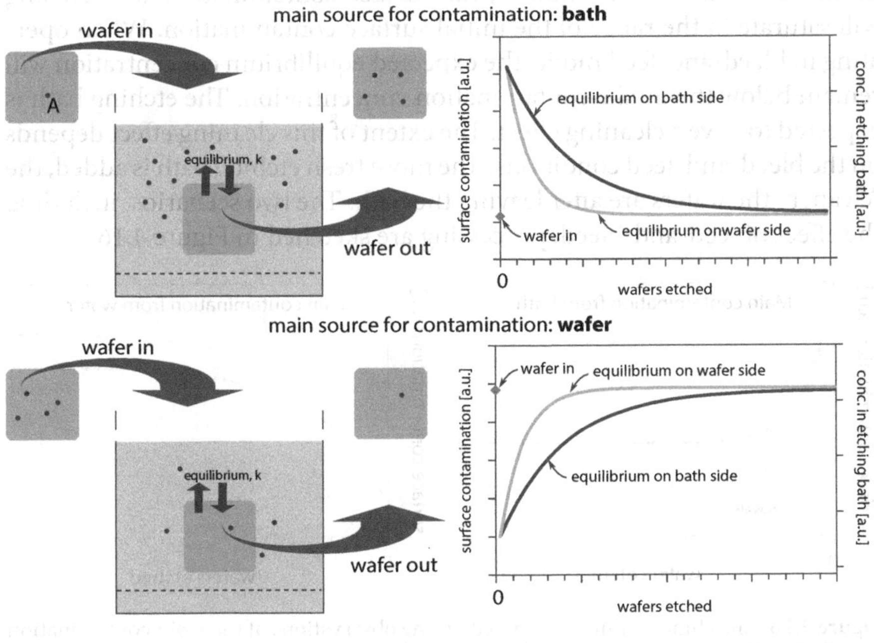

Figure 4.15 In this figure, two cases are discriminated. Top: main source etching bath; bottom: main source the wafer.

equilibrium determines how fast the bath and wafer surface content will run into saturation. The main source of the contamination (bath or wafer) determines the general shape of the curve. In both cases, the equilibrium wafer contamination level after etching is expected to be the same as before etching: $A_{surface} \approx A_{surface,i}$.

Due to potential losses through precipitation and sedimentation, this might not quite hold true. Yet, the trends in Figure 4.14 indicate good agreement with this theory. The main sources for Fe, Ni, and Mo is the etching bath. It can be concluded that the initial as-cut surface concentration of Ni is reached after seven batches. Fe and Mo are expected to reach the point of equilibrium roughly after tens or several tens of batches.

This is the case, when no fresh chemicals are added. In mass production, however, "fresh" contaminants are added under feed-and-bleed conditions. Consequently, the equilibrium concentrations on the wafers exiting the etching bath will shift toward higher surface contamination and become mostly driven by the quality of the etching bath and the feed-and-bleed amount.

The main source for Cu (and also Ti and Cr), on the other hand, is the as-cut wafers. Without feed-and-bleed, the wafer surface contamination is firstly expected to be removed from the wafer surface and thus to be enriched in the bath. Eventually, the surface contamination after etching will saturate in the range of the initial surface contamination. When operating in bleed-and-feed mode, the expected equilibrium concentration will remain below the intake contamination concentration. The etching bath is expected to have a cleaning effect. The extent of this cleaning effect depends on the bleed-and-feed conditions. The more fresh etching bath is added, the "cleaner" the wafers are after leaving the bath. The two scenarios, including the effect of feed-and-bleed, processing are sketched in Figure 4.16.

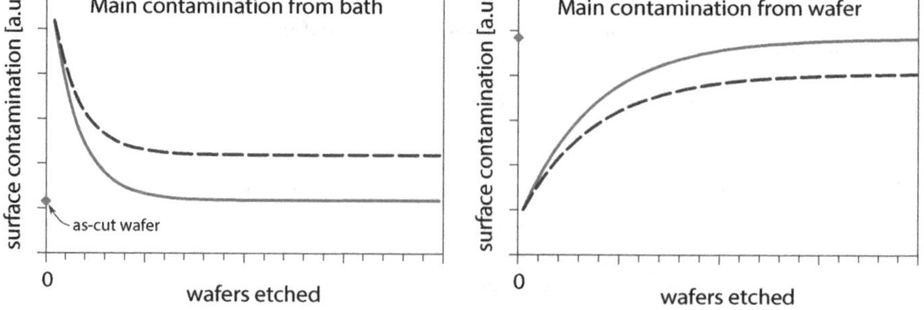

Figure 4.16 Enrichment simulation based on the observations of the main contamination sources being the etching bath (left) or the wafer (right), full lines without and dashed lines with feed-and-bleed processing [56].

The behavior of Ti and Cr is indicating that the dynamics between surface contamination, bath age, and the respective etching bath concentrations these cases may be more complicated. Both elements seem to be removed increasingly effectively for the first roughly four batches of wafers. Then, the surface contamination appears to be stabilizing (Cr) or the element starts enriching again (Ti). Ti is expected to enrich in the bath until "the intake equals output" point is reached. More likely, the Cr surface contamination after etching will eventually start to increase, as Cr is enriched in the etching bath. One can only speculate about the initial increase in removal efficiency of both elements. One of the reasons may be changes in the etching bath, such as the consumption of NaOH or the increase in Si concentration and thus connected an initial increase in solubility.

One of the major implications of this behavior is that, if instead of as-cut wafers at the front end of the line (FEOL) in the process sequence, clean wafers are textured in a later process step (BEOL), which may be the case in alternative PERC process sequences, e.g., as published by Münzer *et al.* [63] or for PERT or IBC solar cells, the etching bath is the sole contamination source and will be "cleaned" by the wafers. This assumption is confirmed by the data shown in Figure 4.17 [56].

Indeed, the surface contamination values are much lower in the BEOL case, and the values are expected to decrease over time, depending on the feed-and-bleed conditions. Especially when fresh baths are used in the BEOL case, relatively high metal concentrations (1e+11 to 1e+14cm^{-2}) are detected, stemming from the alkaline texturing solution. These values are backed by findings on HF/HNO_3 textured wafers, in which similar concentrations were found to be emitted by the diluted KOH (2% wt.) rinsing step, which is necessary to remove the porous silicon [31]. As the values are dominated by the etching bath, the surface concentrations after etching are expected to

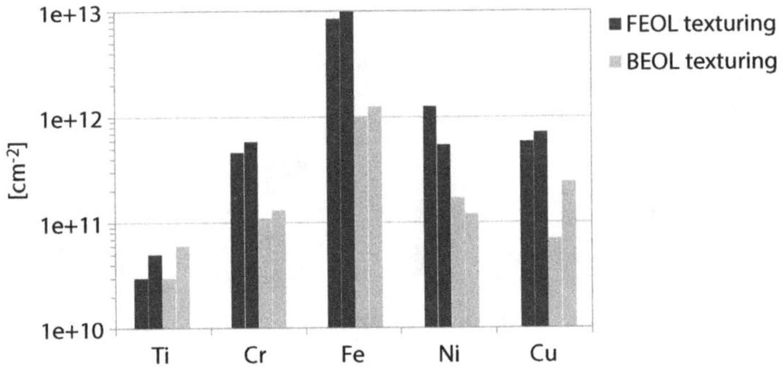

Figure 4.17 Surface contamination after alkaline texturing: of as-cut wafers (FEOL), and during the process (BEOL). Two pairs of samples were measured for each condition [56].

differ with the kind and grade of the used chemicals. If KOH, for example, is replaced by high quality TMAH, a common alkaline etchant in the semiconductor industry, the subsequent cleaning may be strongly simplified. This, however, is hardly feasible due to the high costs of the chemicals.

4.4.4 Contaminants of Interest and Their Threshold Values

After having established the contamination levels observed once the wafers have been treated by etching solution, the threshold values that are needed after cleaning would be of interest for the practical use of this type of analysis. As surface contamination from sawing or etching is made up from numerous species, determination of thresholds for each one would be tedious work, especially when considering that different process steps may have different sensitivities. Some values for single elements have been published [35, 64], yet for practical application—namely, the use of surface or solution contamination data—good knowledge of a small number of contaminants might be useful.

As elaborated in the previous section, highest surface concentrations of metals are found on as-cut wafers and the initial surface contamination is hardly reduced by the first alkaline etching step. An example for the contamination load of diamond wire sawed Cz wafers after alkaline etching is given in Figure 4.18. The high Fe and Cu concentrations on the surface are stemming from the as cut wafer (cf. Section 4.4.2) and the alkaline etching baths (cf. Section 4.4.3).

As shown above, Fe and Cu are the most common contaminants in a process surrounding. Especially Fe, as main component of steel, can easily

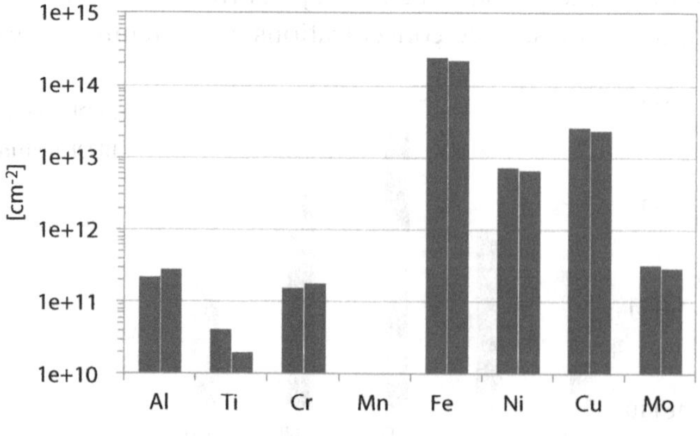

Figure 4.18 Example for surface contamination of diamond-wire sawed Cz-wafers after saw damage etch in NaOH (22%), as measured by sandwich etch ICP-MS, two neighboring sets of wafers were analyzed. Two pairs of samples were measured [56].

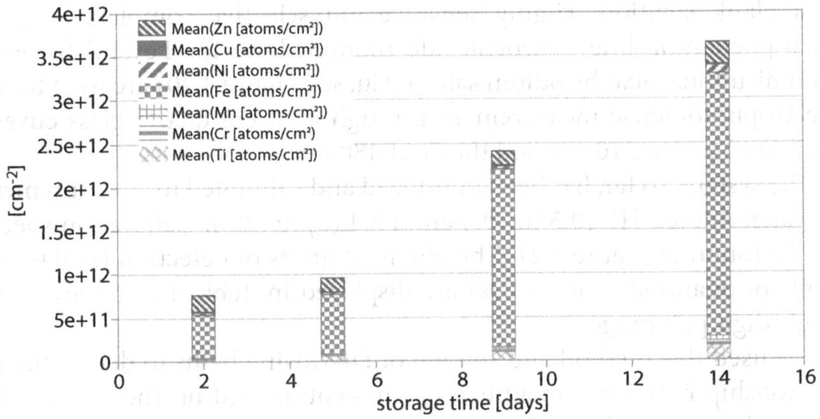

Figure 4.19 Metal contamination collected on wafer surfaces stored over time in carriers in a solar cell fabrication. Wafers surfaces were extracted after storage using the droplet sandwich etch method. Data has been combined from several measuring points. Order of elements according to list from top to bottom. Unpublished data from the "5ct" project.

be transferred to the wafer surface, as can be seen in Figure 4.19 either by metal particles contained in dust—solar cell fabrication often does not employ clean room environment—or in direct contact with metal surfaces (measurement chucks, conveyor belts) or by contaminated water or etching/cleaning baths. Hence, Fe may be considered as representative of contamination originating from the fabrication surrounding.[3]

Naturally, other elements will find their ways on the wafer surface. However, insufficient cleaning, on the one hand, will most of the times result in increased Cu contamination, as Cu, due to its noble nature, is easily reduced to the elemental state on the wafer surface [65]. Thus, Cu can be used as marker for insufficient cleaning.

Consequently, these two elements may be used as markers for the detection and monitoring of metal contamination—the contamination of interest. Accordingly, we have developed a straight forward on-site characterization device that can measure those two elements in aqueous (acidic) solutions such as cleaning and rinsing baths. The method is based on ion selective complexing agents and spectrophotometric analysis [56, 64].

The whole necessary sample preparation procedure (mixing, pH adjustment, etc.) is run by a liquid handling system. It controls the modules and data handling, processes, and stores the collected data. The calibration of the system is performed by adding defined amounts of a Fe/Cu mixture

[3]The data was gathered in the "5ct" research project, funded by the German Ministry for Economic Affairs (0324092C).

to a blank solution. Highly sensitive ion selective complexing agents (1,10-phenanthroline hydrochloride monohydrate for Fe and bathocuproindisulfonic acid bisodium salt for Cu, see Figure 4.20) are used for the spectrophotometric measurement through a 10-cm quartz glass cuvette. Cu is analyzed at 510 nm and the Fe at 480 nm.

The system, so far, has been optimized and calibrated for measurements of water, diluted HF (0.5 to 2%) and HCl (up to 3%). Calibration spectra can be found in Figure 4.21. The obtained limits of detection (LODs) and limits of quantification (LOQs) are displayed in Table 4.1 and are in the single digit ppb range.

We used this method and spiking of DI-rinsing baths to determine the relationship between contamination in solution and on the surface. The wafers that were subjected to those spiked baths were processed to symmetrical lifetime samples using common industrial process equipment, such as oxidation and diffusion furnaces or PECVD deposition chambers in order to obtain critical contamination values. The lifetime was measured

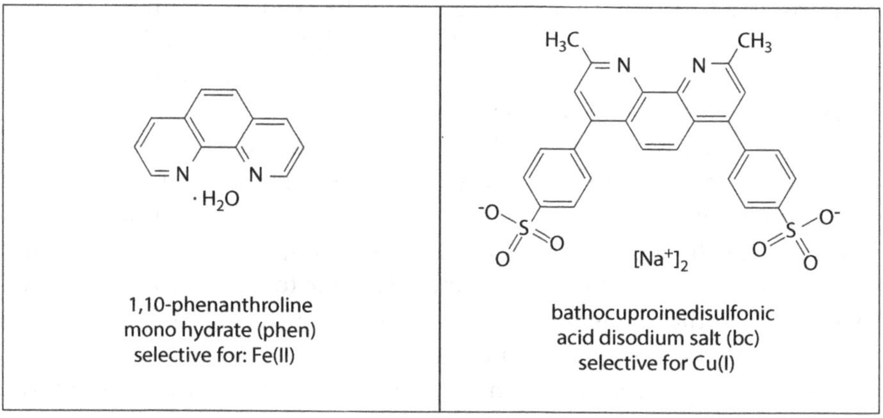

Figure 4.20 Ion selective complexing agents used for the determination of Fe and Cu. For more details, refer to [64].

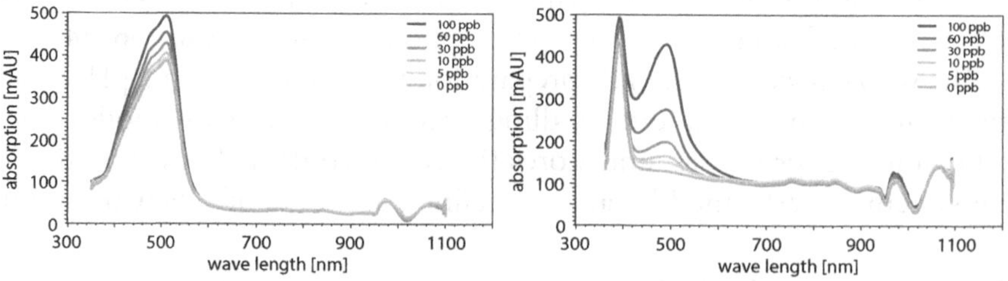

Figure 4.21 Calibration spectra of Fe (left) and Cu (right) in 2% HF [64].

Table 4.1 Detection and quantification limits calculated according to [66] of the METROHM analysis system [64].

	Fe in H_2O	Cu in H_2O	Fe in HF (2%)	Cu in HF (2%)	Fe in HCl (3%)	Cu in HCl (3%)
Limit of detection (LOD) [ppb]	2.2	2.2	2.7	1.1	0.5	6.6
Limit of quantification (LOQ) [ppb]	7.0	7.0	9.5	3.8	1.6	23

and correlated with the surface contamination (and the metal content in the DI water). For the latter relationship, refer to [56, 67]. For threshold values, a loss in lifetime of 1%rel. was defined. The exponential and Gompertz fits were used to extrapolate the data. An example is shown in Figure 4.22. There, an all PECVD-passivation layer (SiOxNy based) and a thermal oxide were compared. The behavior of the first in respect to lifetime is expected to be similar to today's state of the art AlOx passivation stacks, as the thermal budget is very similar. The threshold values for the p-type samples in this experiment are in a very low range for Fe and in a significantly higher range for Cu.

In general, we found for both n-and p-type substrates critical Fe contamination values in the range of 4–10e+10cm^{-2} and for Cu in a similar range for n-type, but significantly higher for p-type substrates. In particular, the $POCl_3$ diffusion process is highly effective in "gettering" and deactivating many contaminants—one reason why Al-BSF solar cells could be produced with very low cost cleaning procedures. In general, the sensitivity depends

Figure 4.22 Correlation between surface contamination and implied Voc for PERC rear side passivation. SiOxNy passivation refers to a PECVD-deposited layer and thermal oxidation to a thermal oxide layer of 10 nm. Both types of samples were fired after SiNx capping. Threshold ranges are indicated in blue. The gray symbols indicate values below the level of quantification of the DSE-ICP-MS. Data from [56].

on the process steps and the solar cell concept. For more details, refer to [56, 67]. Critical concentrations in rinsing water for the above tested rear side passivation stacks were found to be between 40 and 50 ppm. These values can easily be measured with the above shown method.

4.4.5 Organic Contamination

Apart from inorganic (metal) contamination, organic contamination on the wafer surface may also occur. Some examples for the surface coverage by organic content ("carbon surface concentration" or total organic carbon, TOC) are given in Figure 4.23. The left-hand side graph shows the organic contamination after several wet process steps. Alkaline texturing proves to be the strongest source for organic surface contamination, which can be explained by the organic surfactant added to the texturing solution that adheres to the surface even after extensive rinsing. However, also after saw damage etching increased carbon concentrations were measured. Both cleaning sequences after the alkaline etching steps prove to reduce the carbon concentration significantly. In the right-hand side graph, surfaces after texturing were investigated. The first bar indicates the carbon concentration after texturing, the three subsequent bars after post-texture cleaning. Both, pSC1 and the ozone-based cleaning seem to effectively remove the contamination, while the chlorine-based cleaning solution adds carbon to the surface. The latter can be explained by the use of graphite electrodes for the electrolysis of

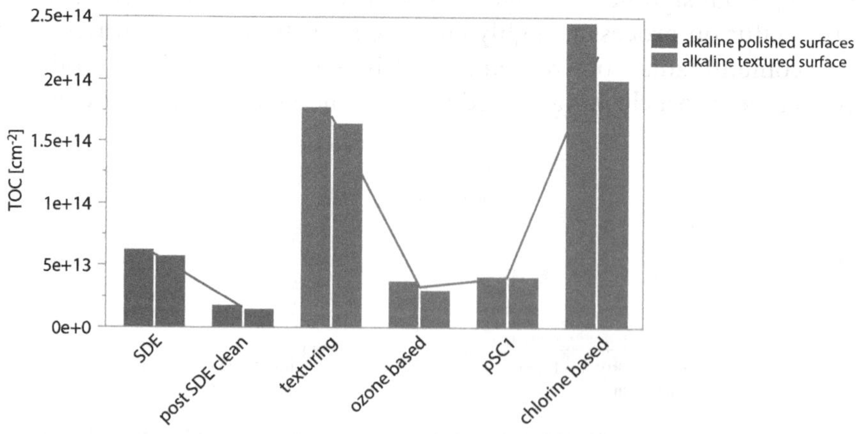

Figure 4.23 Carbon concentration per surface area, two pairs of measurement results from the left (in blue): after SDE and post-SDE clean; four pairs of results from the right (in red): textured surface and three different post-texture cleaning. Unpublished data from the 5ct project.

diluted HCl in order to enrich the solution with chlorine. It appears as if some carbon dissolves into the solution and reaches the wafer surface.

The question of how the organic contamination interferes with the solar cell processing is difficult to answer. If the wafers are subjected to high temperature processes in oxygen containing atmosphere, one can expect that organic residues will simply be combusted to CO_2 and H_2O. However, especially before alkaline etching, organic residues have been known to impact the etching quality [29, 68, 69] by creating stains on the wafer surface, as the effect of the organic species at the surface interfere with the surfactant (additive) of the alkaline texturing solution. Potential sources for such stains are insufficient cleaning of wafers after the wire sawing process, leaving slurry or glues traces behind. If such traces are observed frequently, pre-cleaning, e.g., by pSC1 solution might be helpful [29]. Fingerprints, by bare hands or by nitrile gloves can be detected sometimes on the wafer surfaces by position sensitive lifetime measurement (such as photoluminescence technique) [70] or even with the naked eye. In the solar cell production line, these fingerprints occur when manual handling of wafers becomes necessary for some reason—e.g., when robots fail and wafers have to be saved manually. If in these cases the visible traces are caused by organic contamination or by damaging of the surface or by metal or particles dissipated by the gloves is not clear.

4.4.6 Application: Optimization of HF/O_3 Cleaning Process

State-of-the-art texturing additives for alkaline texturing tend to bond so effectively to the silicon surface that even long rinsing times and dipping in diluted HCl/HF do not remove them completely, which can easily be detected as the surface remains hydrophilic (and as measured and shown in the previous section).

The remaining film may function as mask preventing the cleaning solutions to effectively remove the metal contamination off the surface or interfere with the subsequent process step. In order to remove the remains of the additive that render the surface hydrophilic, an oxidative cleaning step is required, with pSC1 being the reference process.

In order to optimize this process with regard to find the optimum cleaning conditions of the mixture, a DOE using the SAS software JMP10 [71] was carried out. For more details, refer to [56]. The samples were processed according to the above mentioned indirect method of surface preparation for implied Voc measurement (cf. Section 4.4.1). The varied factors were wafer sawing process, HF concentration in the ozone-fed bath and cleaning time. The varied factors can be found in Table 4.2.

Table 4.2 DOE for the cleaning of textured wafers (slurry/diamond wire sawed).

Group #	Sawing process	c(HF) [%]	Time [min]
1	dia	0.3	5
2	slurry	0.3	5
3	slurry	0.3	10
4	slurry	0.3	10
5	dia	1	5
6	slurry	1	2
7	dia	1	10
8	dia	0.01	2
9	slurry	0.01	5
10	dia	0.01	10
11	dia	0.03	10
12	slurry	0.03	2
13	dia	0.03	5
14	dia	0.1	2
15	slurry	0.1	2
16	slurry	0.1	10
17	slurry	0.01	10
18	dia	0.1	10
19	dia	0.01	10

The reference batch (REF) was cleaned with a SPM+HF step after the pSC1+HF/HCl, in order to obtain maximum cleanliness.

The statistical model in Figure 4.24 was obtained by stepwise testing all the possible combinations of factors from first to second order (A, B, A*A, A*B, ...) and for the single factors additional third order influences (A^4, B^3, ...) for statistical relevance. The regression was performed stepwise until the minimum

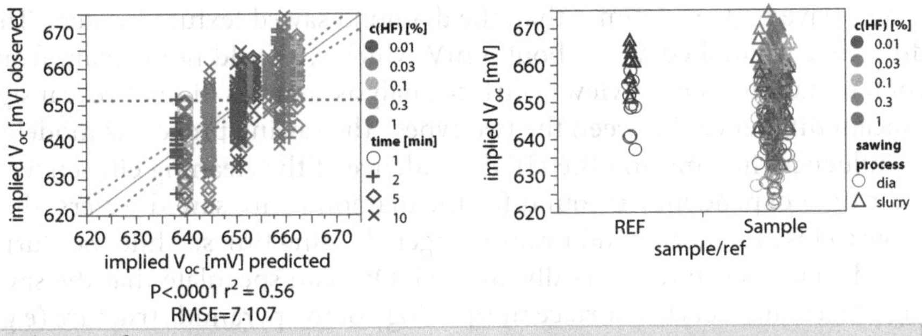

Estimates of parameters

Term	Estimates	Std.-dev.	t-value	Likelihood>|t|
Intercept	649.79	1.923	337.93	<.0001*
time [min]	0.4149	0.126	3.29	0.0011*
sawing process[dia]	-5.842	0.329	-17.77	<.0001*
-log(c(HF))	5.784	1.60	3.61	0.0003*
(time [min]-7.036)*(time [min]-7.036)	-0.1622	0.048	-3.38	0.0008*
(time [min]-7.036)*sawing process[dia]	0.7101	0.096	7.38	<.0001*
sawing process[dia]*(-log(c(HF))-1.029)	3.201	0.421	7.60	<.0001*
(-log(c(HF))-1.029)*(-log(c(HF))-1.029)	-8.012	0.845	-9.48	<.0001*
(-log(c(HF))-1.029)*(-log(c(HF))-1.029)*(-log(c(HF))-1.029)	-6.804	1.730	-3.93	<.0001*

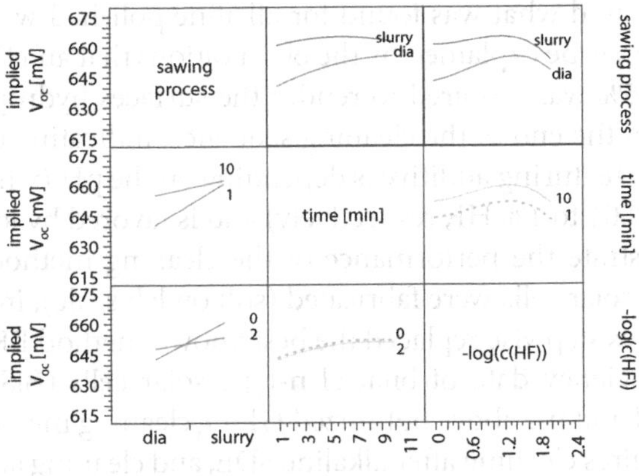

Figure 4.24 Statistical model for the cleaning efficiency of different HF/O_3 cleaning systems on differently sawed wafers after alkaline texturing. Left graph top: Test of model against observed data. Right graph top: Comparison of test samples against the SPM reference to check for the influence of material properties. Mid: Parameter estimates of the model. Bottom: Interaction profiles of the model. Taken from [56].

Bayesian Information Criterion (BIC) was reached.[4] In order to obtain best fits, the common logarithm for the HF concentration (in %) is used.

The first deduction from the regression model, backed by comparison of the references in the top right graph, is that the slurry sawed textured

[4]The BIC is defined as −2 log(likelihood)+k ln(n) with −2 log(likelihood) = n∗(ln(2∗π∗sse/n)+1); sse is the sum of squares for the current model, k is the number of parameters and n is the sample size [71].

wafers have higher lifetimes than the diamond sawed textured wafers. The difference in implied V_{oc} is about 5 mV which we could not explain from contamination point of view—surface analysis samples did not reveal significant differences between the two types. The sawing process is modeled to influence the time and the HF dependence of the cleaning efficiency. A clear time dependency is found for the diamond wire sawed wafers: they show a close to linear trend toward longer cleaning times, while the slurry sawed wafers seem to be hardly affected. One can speculate that the sawing process impacts the surface morphology of the pyramid structure (e.g., due to the sawing marks), for the diamond wire sawing case in a negative way. HF/O_3 has a very small etching rate—yet, if the surface is particularly "rough" as could be in the diamond wire case (which would explain the lower implied V_{oc} level), it could be measurably smoothing the surface with longer treatment time and improving the implied V_{oc}.

The optima for HF concentrations are found between $-\log(c(HF)) = 1.2$ and 1.4 corresponding to 0.06% and 0.04%, which are slightly higher concentrations compared what was found for alkaline polished wafers: 0.02% and 0.025%. This may be explained by the observations that an HF concentration of at least 0.05% was required to render the surfaces hydrophobic after the HF/HCl dip at the end of the cleaning sequence, indicating oxidation of the remains of the texturing additive is depending on the pH (which varied from 2.7 to 1.7 for 0.01 to 1% HF, respectively) and is favored by lower pH values.

To demonstrate the performance of the cleaning methods on solar cell performance, solar cells were fabricated (still on lab scale), in which the new cleaning recipes stepwise replaced the best known method BKM. Figure 4.25 shows the efficiency data of bifacial n-type solar cells (BiSoN), processed with the BKM and the above optimized HF/O_3 cleaning method. The process sequence requires cleaning after alkaline SDE, and cleaning after alkaline texturing. The data indicates that for this solar cell type, the used material, and the given reached efficiency, HF/O_3 indeed can be used as an alternative for the BKM (the overlapping circles in the graph indicate no statistical difference between the groups). As a matter of fact, the used recipes have by now proven their reliability and performance in industrial manufacturing to date.

4.5 Cost Considerations

This last section is dedicated to the costs attributed to wet chemical processing and potential cost savings in particular. The impact of using ozone-based cleaning solutions on the reduction of the running costs will be discussed in more detail.

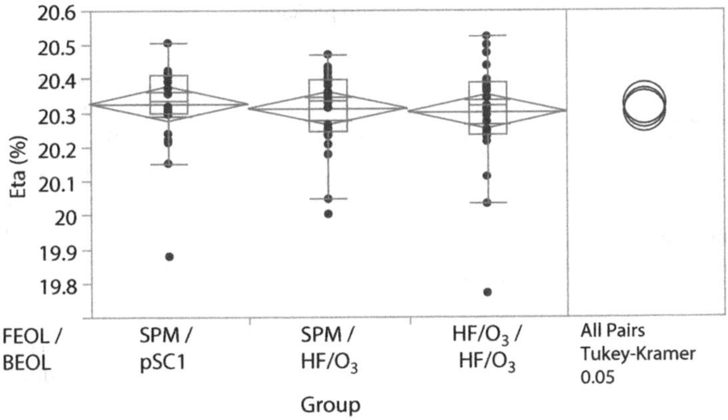

Figure 4.25 Solar cell efficiencies on bifacial n-type solar cells (BiSoN). The BKM (left) was stepwise replaced with HF/O_3-based cleaning recipes. The FOEL cleaning (HF/O_3, SPM) refers to cleaning after SDE and the BEOL cleaning (HF/O_3, pSC1) refers to cleaning after alkaline texturing. Unpublished data from internal quality assurance measurements.

An overview on cost distribution of the process steps and consumables required to manufacture a state-of-the-art PERC cell is given in Figure 4.26. The two wet chemical process steps are highlighted. The data for the relative cost structure is obtained from [72] and refers to a greenfield factory for mono PERC solar cell production in urban China in the first half of 2018. The average efficiency of the produced cells is assumed to be 21.5% and the area of the cells to be 244 cm^2. Overall cost for such as solar cells (including costs for facilities) sum up to 7.3 USD cents/W_{peak}, excluding the wafer (the wafer would add another 11.7 USD cents/W_{peak} to the calculation). When combining 60 of the cells to a typical solar module, the cells make up a little less than 50% of the overall costs. The absolute values for the prices strongly depend on the region of the world, where the respective factory would be located and are driven by factors such as financing costs, scaling effects, materials, and the availability of low cost silicon wafers. Yet, drawing from the data from the cited report, we expect that the relative distribution of costs does not depend too heavily on location.

The screen printing takes the largest piece of the pie, which is mainly due to costs of the silver paste, which for the front side metallization has silver content of around 98%. The wet chemical process steps still take one-fourth of the overall cell production costs. Thus, cost savings without compromising the outcome (the efficiency of the final cells) will have a measurable effect on the overall production costs. Of the two wet steps, the step "SDE, texturing and cleaning" in Figure 4.26 with 21% takes the largest piece and thus is most interesting in respect of cost reduction and

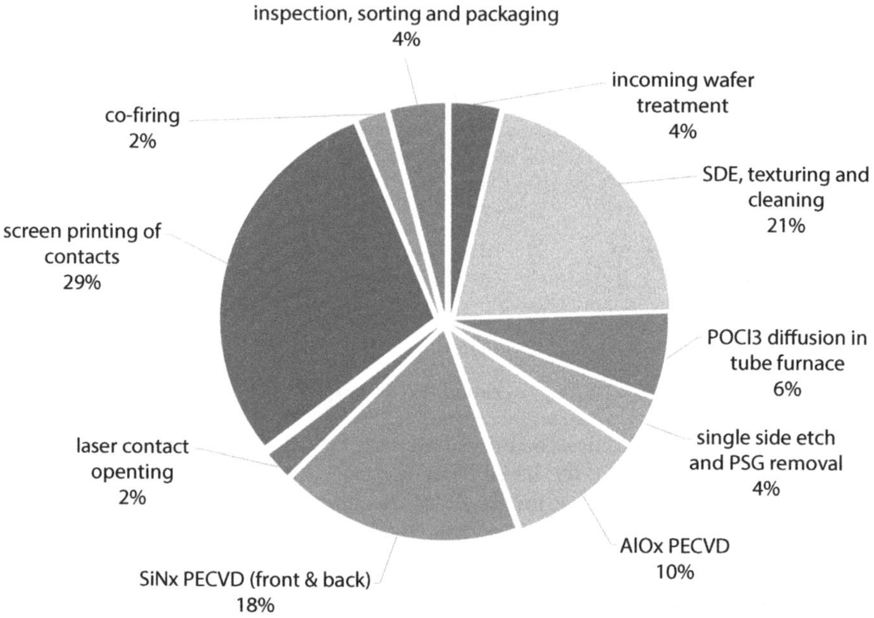

Figure 4.26 Relative process related costs (derived from [72]). It takes into account: consumables, labor, electricity, maintenance, depreciation of equipment and facilities, and R&D and SG&A expenses.

will be discussed in more details. The first possibility for cost reduction is to reduce the etching depth, as it directly scales with the consumption of KOH or NaOH, and size of the tools (short immersion times require less process baths). Due to improving quality of the incoming wafer material (decreased saw damage depth due to the use of thinner wafers), the separate saw damage removal step has been widely abandoned. Accordingly, we will leave it out in the following.

As mentioned in Section 4.3.3, hydrogen peroxide makes up about 50% of the overall costs of the chemical consumption of a typical texturing wetbench, where it is typically used for pre- and post-texture cleaning. A more detailed cost breakdown of such a tool is given as A in Figure 4.27. It assumes a gross throughput of 8,000 wafers per hour. Depreciation is calculated with five years (linearly). In A to C, a downtime of 6.4% was assumed (about 1.5 hours in a 24/7 production line for bath change and maintenance). The chemicals make up the "supplies" part of the chart. The CAPEX ("invest") can only be attributed to about 7% (a value that to a certain degree depends on the supplier, the number of ordered tools, and other factors). Significantly larger amounts have to be spent on utilities (electricity, water, heat, etc.) and on the disposal of the chemical waste.

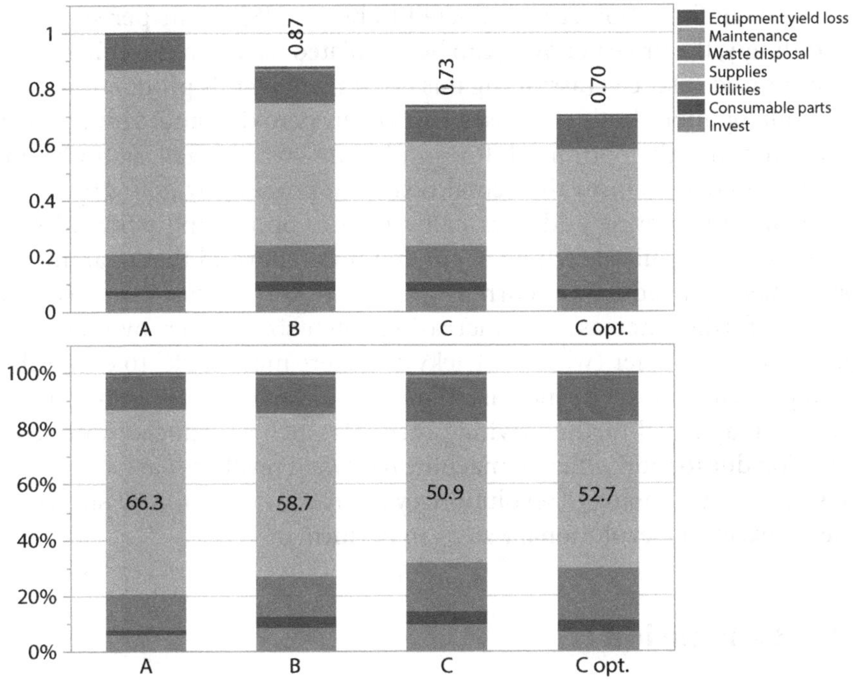

Figure 3.27 Cost break-down of a wet chemical process tools. Throughput of the tools, 8,000 wafers per hour. Tool A: H_2O_2 used for pre- and post-clean after texturing. Tool B: Ozone cleaning is replacing one of the two baths (post-clean). Tool C: Wetbench without H_2O_2. Tool C opt. refers to a fully optimized system (in terms of invest) and reduced downtime.

Both sum up to about 12%. Yield loss, maintenance, and spare parts take negligible percentages in the overall cost calculation.

Accordingly, the largest potential for cost savings is replacing the hydrogen peroxide by an alternative oxidizing agent, which is done in the scenarios B and C. Stepwise one of the two hydrogen peroxide rinses is replaced by an ozone-fed cleaning bath. The ozone generator drives the increase in invest from A to B. Yet, as only one generator is necessary to feed several process baths, only little further invest is necessary when removing hydrogen peroxide from the tool completely. It is to say that version B can be currently referred to as common practice at the moment. When moving completely to ozone-fed baths, the supply costs drop significantly. An overall benefit of up 30% is deemed possible, which is reached in the C opt. tool setup. This option is derived from experience that some more savings are possible when the hardware setup is reduced to the absolutely necessary, also the scheduled downtime was reduced for this tool setup by 3.4%, which can be achieve by increased cycle times. The assumed price for the reference configuration A is estimated to be 2.2 USD cents per substrate. Accordingly, the price per W_{peak} can be calculated based on the efficiency of the produced solar cell (assuming it is not limited by this production step).

In our opinion, both processes (hydrogen peroxide and ozone based) have been mostly optimized for minimum costs. Small optimization potential may remain in the actual texturing process, by going to higher temperature-tolerating additives or by further optimizing pyramid sizes (to reduce the required etching depth and time). Beyond these parameters for further cost reduction, currently, only the throughput remains. This could be further increased by increasing batch sizes, either by increasing the wafers per carrier (which is tricky as wafers may "stick" to each other when packed to close) or increasing the amount of carriers per batch. The scale-up may entail further saving potentials, such as reduced water consumption due to more efficient machine hardware (evolutionary approach). New process technology (revolutionary approach) can only be successful when it excels the evolutionary steps in performance.

4.6 Conclusion

In this chapter, a look at etching and cleaning technology from an industrial point of view is investigated. The most commonly used process technology—batch- and inline-type tools—were introduced and explained.

Clean wafer surfaces are required, particularly with solar cell efficiencies reaching higher values. In order to run a high-efficiency solar cell production line to date and in the future, we emphasize the importance of a deeper knowledge of contamination sources and process variation (in particular of alkaline processes). We have shown that the main sources for process-induced metal contamination are as-cut wafers (to a varying degree) and alkaline etching solutions. The second type of contamination that was addressed is organic contamination that can interfere with etching and cleaning steps. In order to produce sufficiently clean enough surfaces—to remove noble elements and organics—oxidative cleaning solutions are required. To date, hydrogen peroxide base solutions are most commonly used.

Yet, wet chemical process steps make up about 25% of the overall production costs, as shown in a cost structure of a PERC solar cell process, of this, the use of hydrogen peroxide makes up significant amount of consumables supply costs. As the industry is constantly driven by reduced production prices, we show that ozone-based cleaning technology can significantly reduce the supply costs easily overcompensating for the higher

CAPEX. Using the indirect methods for determining the surface cleanliness, such alternative cleaning processes can be optimized, which was shown for the example of the HF/O_3 system, which features the additional benefit of producing very lowly concentrated chemical waste.

For the constant monitoring of cleaning and rinsing baths for potential unexpected events of contamination (such as machine failure), the photometric measurement of Fe and Cu as "elements of interest" is suggested. The introduced measurement is highly sensitive and comes at much lower cost than conventional mass spectroscopic methods for trace element analysis and is thus suggested be used for continuous process control to ensure reliable high-yield production.

As the solar industry is striving for further cost reduction, we could show that ozone-based cleaning processes are superior in cost structure. The much lower running cost over compensate the higher invest by far. Using the same type of equipment, we expect up to 30% reduction of production costs for the wet chemical production steps. Potential for further reduction can be found by improving the efficiency of the tool, mainly by increasing the throughput, yet this entails either changing the carrier design or increasing the batch size. The latter may require major rework of the current machine layouts. The already very low costs per substrate of less than 2 USD cents per wafer for a batch texture and cleaning process, indicates with what any new "revolutionary" technology will have to compete.

Acknowledgments

The authors would like to thank the colleagues from ISC Konstanz, for their contribution to the results, Prof. Dr. Kroke for his supervision of the PhD thesis, and the German Ministry for Economic Affairs (BMWi) for funding part of the presented works (5ct, Förderkennzeichen 0324092C).

References

1. Fu, R., Feldman, D.J., Margolis, R.M.U.S., Solar Photovoltaic System Cost Benchmark: 2018, in: *Technical report, National Renewable Energy Laboratory*, vol. 11, 2018. Accessed September, 13th 2019.
2. Trube, J., *International Technology Roadmap for Photovoltaic (ITRPV)*, Tenth Edition, VDMA e.V., Frankfurt, https://itrpv.vdma.org/, March 2019. Accessed June, 21st 2019.
3. Kavlak, G., McNerney, J., Trancik, J.E., Evaluating the causes of cost reduction in photovoltaic modules. *Energy Policy*, 123, 700–710, 2018.

4. Awate, S., Ajith, V., Ajwani-Ramchandani, R., Catch-up as a Survival Strategy in the Solar Power Industry. *J. Int. Manag.*, 24, 2, 179–194, 2018.
5. Fertig, F., Lantzsch, R., Mohr, A., Schaper, M., Bartzsch, M., Wissen, D., Kersten, F., Mette, A., Peters, S., Eidner, A., Mass production of p-type Cz silicon solar cells approaching average stable conversion efficiencies of 22%. *Energy Procedia*, 124, 338–345, 2017.
6. Green, M.A., Photovoltaic technology and visions for the future. *Prog. Energy*, 1, 1, 013001, 2019.
7. Macdonald, D.H., Cuevas, A., Kerr, M.J., Samundsett, C., Ruby, D., Winderbaum, S., Leo, A., Texturing industrial multicrystalline silicon solar cells. *Sol. Energy*, 76, 1–3, 277–283, 2004.
8. Melnyk, I., Wefringhaus, E., Delahaye, F., Vilsmeier, G., Mahler, W., Fath, P., High throughput in-line acidic texturisation and edge isolation: an industrial reality. *Proceedings of the 20th European Photovoltaic Solar Energy Conference and Exhibition*, Barcelona, Spain, pp. 1403–1406, 2005.
9. Kumar, A. and Melkote, S.N., Diamond wire sawing of solar silicon wafers: A sustainable manufacturing alternative to loose abrasive slurry sawing. *Procedia Manuf.*, 21, 549–566, 2018.
10. Meinel, B., Koschwitz, T., Acker, J., Textural development of SiC, diamond wire sawed sc-silicon wafer. *Energy Procedia*, 27, 330–336, 2012.
11. Lippold, M., Buchholz, F., Gondek, C., Honeit, F., Wefringhaus, E., Kroke, E., Texturing of SiC-slurry and diamond wire sawn silicon wafers by HF–HNO3–H2SO4 mixtures. *Sol. Energy Mater. Sol. Cells*, 127, 104–110, 2014.
12. Zhuang, Y.F., Zhong, S.H., Huang, Z.G., Shen, W.Z., Versatile strategies for improving the performance of diamond wire sawn mc-Si solar cells. *Sol. Energy Mater. Sol. Cells*, 153, 18–24, 2016.
13. Jooß, W., Melnyk, I., Lai, C., Jung-König, J., Teppe, A., Madugula, S.R., Voigt, O., Fath, P., Hu, B., Wang, I., Process development for MCCE textured solar cells on multicrystalline DWS wafer. *Proceedings of the 2018 IEEE 7th World Conference on Photovoltaic Energy Conversion (WCPEC)(A Joint Conference of 45th IEEE PVSC, 28th PVSEC & 34th EU PVSEC)*, IEEE, pp. 2783–2787, 2018.
14. Lamb, W.R. and Lawrence, J.E., Method of forming light-trapping surface for photovoltaic cell and resulting structure. U.S. Patent No. 4,918,030. Washington, DC: U.S. Patent and Trademark Office, 1990.
15. King, D.L. and Buck, M.E., Experimental optimization of an anisotropic etching process for random texturization of silicon solar cells. *Proc. 22nd IEEE Photovoltaic Specialists Conference*, Las Vegas, vol. 1, pp. 303–308, 1991.
16. Seidel, H., Csepregi, L., Heuberger, A., Baumgärtel, H., Anisotropic etching of crystalline silicon in alkaline solutions I. Orientation dependence and behavior of passivation layers. *J. Electrochem. Soc.*, 137, 11, 3612–3626, 1990.
17. Cornagliotti, E., Uruena, A., Horzel, J., John, J., Tous, L., Hendrickx, D., Prajapati, V., Singh, S.N., Hoyer, R., Delahaye, F., How much rear side polishing is required? A study on the impact of rear side polishing in PERC solar cells. *Proceedings of the 27th European Photovoltaic Solar Energy Conference and Exhibition*, pp. 561–566, 2012.

18. Kranz, C., Wyczanowski, S., Baumann, U., Weise, K., Klein, C., Delahaye, F., Dullweber, T., Brendel, R., Wet chemical polishing for industrial type PERC solar cells. *Energy Procedia*, 38, 243–249, 2013.
19. John, J. and Haslinger, M., Texturing monocrystalline silicon substrates. U.S. Patent No. 9,876,128. Washington, DC: U.S. Patent and Trademark Office, 2018.
20. Bailey, W.L., Coleman, M.G., Harris, C.B., Lesk, I.A., Reduction of reflectivity. U.S. Patent No. 4,137,123. Washington, DC: U.S. Patent and Trademark Office, 1979.
21. Singh, P.K., Kumar, R., Lal, M., Singh, S.N., Das, B.K., Effectiveness of anisotropic etching of silicon in aqueous alkaline solutions. *Sol. Energy Mater. Sol. Cells*, 70, 1, 103–113, 2001.
22. Vazsonyi, E., De Clercq, K., Einhaus, R., Van Kerschaver, E., Said, K., Poortmans, J., Szlufcik, J., Nijs, J., Improved anisotropic etching process for industrial texturing of silicon solar cells. *Sol. Energy Mater. Sol. Cells*, 57, 2, 179–188, 1999.
23. Hauser, A., Melnyk, I., Wefringhaus, E., Delahaye, F., Vilsmeier, G., Fath, P., Acidic texturisation of mc-Si using a high throughput in-line prototype system with no organic chemistry. *Proceedings of the 19th European Photovoltaic Solar Energy Conference and Exhibition*, Paris, pp. 1094–1097, 2004.
24. Graff, K., *Metal impurities in silicon-device fabrication*, vol. 24, Springer-Verlag, Berlin, Heidelberg, 2000.
25. Davis, J.R., Rohatgi, A., Hopkins, R.H., Blais, P.D., RaiChoudhury, P., McCormick, J.R., Mollenkopf, H.C., Impurities in silicon solar cells. *IEEE Trans. Electron Devices*, 27, 4, 677–687, 1980.
26. Zhong, L. and Shimura, F., Dependence of lifetime on surface concentration of copper and iron in silicon wafers. *Appl. Phys. Lett.*, 61, 9, 1078–1080, 1992.
27. Istratov, A.A., Buonassisi, T., McDonald, R.J., Smith, A.R., Schindler, R., Rand, J.A., Kalejs, J.P., Weber, E.R., Metal Content of Multicrystalline Silicon for Solar Cells and its Impact on Minority Carrier Diffusion Length. *J. Appl. Phys.*, 94, 6552, 2003.
28. Chen, G. and Kashkoush, I., Effect of pre-cleaning on texturization of c-Si-wafers in a KOH/IPA mixture. *ECS Trans.*, 25, 15, 3–10, 2010.
29. Sonner, C., Oltersdorf, A., Zimmer, M., Rentsch, J., Influence of Contaminations and Cleaning Sequences on Alkaline Texturisation. *Proceedings of the 26th European PV Solar Energy Conference and Exhibition*, Hamburg, Germany, 2011.
30. Moldovan, A., Feldmann, F., Krugel, G., Zimmer, Ma., Rentsch, J., Hermle, M., Roth-Fölsch, A., Kaufmann, K., Hagendorf, C., Simple cleaning and conditioning of silicon surfaces with UV/ozone sources. *Energy Procedia*, 55, 834–844, 2014.
31. Buchholz, F., Wefringhaus, E., Eisert, S., Egenolf, H., Groß, A., Measurement and Impact of Surface Transition Metal Contamination of Textured

Multicrystalline Silicon Wafers. *Proceedings of the 26th European Photovoltaic Solar Energy Conference and Exhibition*, Hamburg, pp. 1187–1190, 2011.
32. Wostyn, K., Baekelant, W., Rip, J., Haslinger, M., Kenis, K., Struyf, H., Claes, M., Mertens, P.W., De Gendt, S., "Just clean enough": wet cleaning for solar cell manufacturing applications, in: *Solid State Phenomena*, vol. 195, pp. 293–296, 2013.
33. Feitknecht, W. and Schindler, P., Solubility constants of metal oxides, metal-hydroxides and metal hydroxide salts in aqueous solution. *Pure Appl. Chem.*, 6, 2, 125–206, 1963.
34. Schweckendiek, J., Hoyer, R., Patzig-Klein, S., Delahaye, F., Knoch, G., Nussbaumer, H., Cleaning in Crystalline Si Solar Cell Manufacturing, in: *Solid State Phenomena*, 195, 283–288, *Trans. Tech. Publ.*, 2013.
35. Buchholz, F., Wefringhaus, E., Schubert, G., Metal surface contamination during phosphorus diffusion. *Energy Procedia*, 27, 287–292, 2012.
36. Kern, W., Cleaning solution based on hydrogen peroxide for use in silicon semiconductor technology. *RCA Rev.*, 31, 187–205, 1970.
37. Strinitz, F., El Jaouhari, A., Schörg, F., Fürst, M., Plettig, M., Kühnlein, H., Advanced alkaline texturing and cleaning for PERC and SHJ solar cells. *Energy Procedia*, 130, 23–30, 2017.
38. Bohling, D.A., George, M.A., Goldstein, A., Lizard, T., Treichel, H., Silicon Solar Cell Efficiency Improvement Through Advanced Cleaning Technology, TechConnect Briefs. *Silicon Solar Cell Efficiency Improvement Through Advanced Cleaning Technology*, pp. 80–83, 2010.
39. Nicoll, W.D. and Smith, A.F., Stability of dilute alkaline solutions of hydrogen peroxide. *Ind. Eng. Chem.*, 47, 12, 2548–2554, 1955.
40. Bergman, E. and Lagrange, S., Process and environmental benefits of HF-ozone cleaning chemistry. *Solid State Technology*, 44, 7, 115, 2001.
41. Bergman, E.J., Lagrange, S., Claes, M., De Gendt, S., Röhr, E., Pre-Diffusion cleaning using ozone and HF. *Solid State Phenomena*, 76, 85–88, 2001.
42. Choi, G.M., Yokoi, I., Ohmi, T., The role of oxidant in HF-based solution for noble metal removal from substrate. *Solid State Phenomena*, 76, 267–270, 2001.
43. Choi, G.-M. and Ohmi, T., Removal efficiency of metallic impurities on various substrates in HF-based solutions. *J. Electrochem. Soc.*, 148, 5, G241–G248, 2001.
44. Fukazawa, Y., Miyazaki, H., Ogawa, Y., Direct replacement cleaning technology based on ozonated water using a single processing tank. *Proc. International Symposium on Cleaning Technology in Semiconductor Device Manufacturing*, Paris, Electrochemical Society, pp. 264–271, 1998.
45. Gottschalk, C. and Schweckendiek, J., Using dissolved ozone in semiconductor cleaning applications. *Micro*, 22, 2, 81–84, 2004.
46. Chen, G., The applications of DI-O_3 water on wafer surface preparation. *Proceedings of the International Conference on Wafer Rinsing, Water Reclamation and Environmental Technology for Semiconductor Manufacturing*, Hsinchu, Taiwan, pp. 100–123, 1999.

47. Pipia, F., Bellandi, E., Crivelli, B., Alessandri, M., Fe, and Cu, removal efficiency in HF-DIW/O_3 cleaning sequence. *Sol. State Phenom.*, 65, 109–112, 1998.
48. Moldovan, A., Birmann, K., Rentsch, J., Zimmer, M., Gitte, T., Fittkau, J., Combined ozone/HF/HCl based cleaning and adjusted Emitter etch-back for silicon solar cells. *Sol. State Phenom.*, 195, 305–309, 2013.
49. Chunduri, S.K. and Schmela, M., PERC Solar Cell Technology 2016 - Background, Status and Outlook, Taiyang News, Düsseldorf, April, 4th 2016. http://taiyangnews.info/reports/perc-solarcell-technology-2016/. Accessed June, 21st 2019.
50. Deng, W., Ye, F., Liu, R., Li, Y., Chen, H., Xiong, Z., Yang, Y., Chen, Y., Wang, Y., Altermatt, P.P., 22.61% efficient fully screen printed PERC solar cell. *2017 IEEE 44th Photovoltaic Specialist Conference (PVSC)*, IEEE, pp. 2220–2226, 2017.
51. Altermatt, P.P., Yang, Y., Chen, Y., Fen., Z., Riding the workhorse of the industry: PERC. *Photovoltaics Int.*, 41, 54–64, 2018.
52. Rentsch, J., Ackermann, J., Birmann, K., Furtwängler, H., Haunschild, J., Kästner, G., Neubauer, R., Nievendick, J., Oltersdorf, A., Rein, S., Wet chemical processing for C-Si solar cells status and perspectives. *Proceedings of the 24th European Photovoltaic Solar Energy Conference*, Hamburg, Germany, pp. 1113–1117, 2009.
53. Kashkoush, I., Chen, G., Nemeth, D., Rieker, J., Wafer surface preparation for high-efficiency solar cells. *2014 IEEE 40th Photovoltaic Specialist Conference (PVSC)*, pp. 1211–1213, 2014.
54. Lossen, J., Buchholz, F., Comparotto, C., Eisert, S., J.and Mihailetchi, V.D., Wefringhaus, E., Rossetto, M., Discato, D., Traverso, F., From lab to fab: Bifacial n-type cells entering industrial production. *Proceedings of the 31st European Photovoltaic Solar Energy Conference and Exhibition*, Hamburg, Germany, pp. 965–968, 2015.
55. Buchholz, F., Preis, P., Chu, H., Lossen, J., Wefringhaus, E., Progress in the development of industrial nPERT cells. *Energy Procedia*, 124, 649–656, 2017.
56. Buchholz, F., Metal Surface Contamination in C-Si Solar Cell Processing. PhD thesis, Technische Universität Bergakademie Freiberg. *Fakultät für Chemie und Physik*, 2016.
57. Altermatt, P.P., Yang, Y., Sheng, Y., Chen, D., Chen, Y., Feng, Z., Verlinden, P.J., A method for optimizing PERC cells in industrial production lines using final IV parameters, statistical procedures and numerical device modeling. *AIP Conference Proceedings*, AIP Publishing, vol. 1999, pp. 110001, 2018.
58. Sinton, R.A. and Cuevas, A., Contactless determination of current–voltage characteristics and minority-carrier lifetimes in semiconductors from quasi-steady-state photoconductance data. *Appl. Phys. Lett.*, 69, 17, 2510–2512, 1996.
59. Bowden, S., Yelundur, V., Rohatgi, A., Implied-Voc and Suns-Voc measurements in multicrystalline solar cells. *Conference Record of the 29th IEEE Photovoltaic Specialists Conference*, vol. 2002, IEEE, pp. 371–374, 2002.

60. Wahl, S., Meyer, S., Timmel, S., Quantitative Analysis of Impurities on Semiconductor Materials. *19. Arbeitstagung Angewandte Oberflächenanalytik (AOFA)*, Soest, Germany, 2016.
61. Meyer, S., Wahl, S., Hagendorf, C., Advanced Metal Contamination Analysis for High Efficiency Solar Cell Manufacturing. *Energy Procedia*, 92, 369–373, 2016.
62. Meyer, S., Wahl, S., Timmel, S., Köpge, R., Jang, B.-Y., The impact of wafering on organic and inorganic surface contaminations. *Appl. Surf. Sci.*, 378, 384–387, 2016.
63. Münzer, K.A., Hein, M., Schöne, J., Hanke, M., Teppe, A., Schlosser, R.E., Maier, J., Yodyunyong, A., Krümberg, S., Keller, S., Technical performance, industrial implementation in favour of centaurus technology. *Energy Procedia*, 27, 631–637, 2012.
64. Buchholz, F., Wefringhaus, E., Moron, L., Lukesch, M., Weber, W., Spectrophotometric monitoring of iron and copper contamination of cleaning baths for solar cell processing. *Proceedings of the 28th European Photovoltaics Solar Energy Conference*, Paris, pp. 1813–1815, 2013.
65. Istratov, A.A. and Weber, E.R., Physics of Copper in Silicon. *J. Electrochem. Soc.*, 149, 1, G21–G30, 2002.
66. Kolb, M., Bahr, A., Hippich, S., Schulz, W., Calculation of detection limit, identification limit and determination limit according to DIN32645 with the aid of a computer program. *Acta Hydrochimica et Hydrobiologica*, 21, 6, 308–311, 1993.
67. Buchholz, F. and Wefringhaus, E., Impact of Fe and Cu surface contamination on high efficiency solar cell processes. *Sol. State Phenom.*, 219, 305–311, 2015.
68. Nakamura, K., Aoki, M., Sumita, I., Sato, H., Kumagai, Y., Kawata, Y., Ohshita, Y., Texturization control for fabrication of high efficiency mono crystalline si solar cell. *2013 IEEE 39th Photovoltaic Specialists Conference (PVSC)*, IEEE, pp. 2239–2242, 2013.
69. Park, H., Kwon, S., Lee, J.S., Lim, H.J., Yoon, S., Kim, D., Improvement on surface texturing of single crystalline silicon for solar cells by saw-damage etching using an acidic solution. *Sol. Energy Mater. Sol. Cells*, 93, 10, 1773–1778, 2009.
70. Chen, F.W., Cotter, J.E., Trupke, T., Bardos, R.A., Characterization of PECVD Silicon Nitride Passivation with Photoluminescence Imaging. *Proceedings of the 2006 IEEE 4th World Conference on Photovoltaic Energy Conference*, vol. 2, pp. 1372–1375, 2006.
71. Sall, J., Lehman, A., Stephens, M.L., Creighton, L., *JMP start statistics: a guide to statistics and data analysis using JMP*, vol. 7, SAS Institute Inc., Cary, NC, USA, 2007.
72. Woodhouse, M.A., Smith, B., Ramdas, A., Margolis, Ro. M., Crystalline Silicon Photovoltaic Module Manufacturing Costs and Sustainable Pricing: 1H 2018 Benchmark and Cost Reduction Roadmap, in: *Technical report, National Renewable Energy Laboratory*, vol. 2, 2019. Accessed October, 10th 2019.

5

Analytical Techniques for Wet Processing

Stefanie Wahl* and Sylke Meyer

Fraunhofer Center for Silicon Photovoltaics CSP, Halle (Saale), Germany

Abstract

For current and future solar cell manufacturing processes, control of purity on wafer surfaces and interfaces is essential. Constantly increasing demands for ever greater conversion efficiency on the one hand and ever higher production throughput on the other require ever more specific and sensitive analytical methods. Therefore, this chapter will focus on the advanced chemical analysis of silicon wafer surface contamination. First, a short overview on the impact of surface contamination on the solar cell process is given. Then, this chapter provides detailed description of the analysis of trace metals as well as new developments in analytical techniques for organic contamination. Various analytical techniques are discussed with respect to their applicability for solar wafers and practicability in industrial processes. Especially, high resolution ICP-MS analysis and determination of total organic carbon, subsequent to wet chemical surface extraction, are considered in detail.

***Keywords*:** Surface analysis, sequential etching, metal contamination, organic contamination, surface energy, surface extraction

5.1 Introduction

The development of analytical techniques for the identification and quantification of impurities within the bulk as well as on the surface of semiconductor materials plays an increasing role for the manufacturing of recent photovoltaic modules. This chapter will focus on the chemical analysis of silicon wafer surfaces since the influence of surface and interface layers increases with the maturation of solar cell manufacturing. For the

**Corresponding author*: Stefanie.Wahl@csp.fraunhofer.de

Monika Freunek Müller (ed.) Photovoltaic Manufacturing: Etching, Texturing, and Cleaning, (161–180) © 2021 Scrivener Publishing LLC

same reason, the impact of organic impurities on wafer surfaces increases. Therefore, this chapter provides not only detailed description of the analysis of trace metals but also new developments in analytical techniques for organic contamination. Beyond an overview over recent analytical techniques, examples from application will be given.

5.1.1 The Importance of Chemical Analysis on Silicon Surfaces

After a period of pioneer work in PV industry which has mostly neglected more than very basic chemical analytics, its maturation phase is characterized not least by the introduction of advanced measurements and process monitoring. At a first glance, the enormous cost pressure on the PV industry counteracts this tendency but medium-term higher overall yield, higher cell efficiencies, optimized bath using time, and advanced incoming goods inspection will reduce manufacturing costs and ensure competitiveness. In particular, wafer surfaces have to be monitored more extensively, as it is state of the art in semi-conductor industries.

Due to the peculiarities of the solar wafer manufacturing process, the wafer surfaces are exposed to manifold contamination sources. The wafering process (sawing and subsequent cleaning) plays a key role in determining the purity of the silicon material and the surface contamination on substrates for high efficiency solar cells. In the wafering step, the silicon surface is bare of any oxide layer for the first time. This virgin silicon surface is decorated with metal ions, which can be subsequently incorporated into the freshly formed oxide layer. Depending on the sawing technology (slurry based or diamond wire), there is, first of all, the sawing wire itself which may, depending on its coating, introduce metals like nickel, copper, and iron by abrasion. The next important source of impurities (metals and organics) is the cutting fluid, which is either PEG/SiC slurry or a water based coolant containing surfactants and defoaming agent. Prior to sawing, the brick is mounted onto a beam carrier using special glues. This requires an extra de-gluing treatment followed by a general wet cleaning process. Thus, gluing, de-gluing, and cleaning (surfactants) probably impair the wafer surface purity. Further sources of mainly organic contamination are packaging materials, plastic and lubricants from transport system, and, last but not least, the operators themselves (fingerprints, gloves). To record all those different contaminations is a great challenge since the molecules are not only chemically diverse but also appear unequally as particle, closed film, or single spots. All analytical methods have to consider those specialties.

5.1.2 Impact of Surface Contaminations on the Solar Cell Process

Approaching the theoretical limits in solar cell efficiency, new cell concepts are intensively developed and steadily improved. These high efficiency cell processes not only require very pure silicon material but also increasingly high quality wafer surfaces. Chemical and structural characteristics of the wafer surfaces influence the texturing and passivation process for both concepts, PERC and HIT cells [1, 2]. Metallic wafer contamination can lead to a higher metal contamination in all subsequent processes and may finally result in lower minority carrier lifetimes [3]. A detailed study on the influence of iron and copper contamination on p-type silicon wafers revealed a significant decrease of open circuit voltage V_{OC} at very low surface concentrations for iron but much higher levels for copper [4]. The specific tolerable levels strongly depend on the temperature and time of all processing steps.

On the other hand, it could be shown that not only metallic impurities but also organic residues on the wafer surface may impair down-stream processes such as texturing and vapor deposition steps [5–7]. Since the silicon wafer surface shows strong polarity effects, it is suitable for attracting organic additives containing polar groups. As a result, organic contaminants can get easily adsorbed by the native oxide of the silicon surface [8]. A consequence is, for example, the very often observed emergence of optical defects after pyramid texturization of the wafer surface, as shown in Figure 5.1.

More detailed analysis of the defect surface area revealed an incomplete formation of pyramidal structures within the site of visible defects (Figure 5.2). Hence, different light reflectivity is the reason for the visible inhomogeneity. Empiric studies showed a correlation between the amount of organic

Figure 5.1 Left: As-cut wafer surface without visible defects. Right: The same wafer after KOH/IPA etching (texture bath) with visible spots.

Figure 5.2 SEM top view of wafer surface after alkaline texturing. Left: Reference site with normal reflectivity. Right: Defect site.

surface contamination and the intensity of optical defects after alkaline texturing. Thus, organic impurities can reduce not only texture quality but also the optical yield and light trapping as well as the final cell efficiency. This is, of course, less critical for multi-crystalline silicon wafers which are textured by acidic etching and thereby removing most of the organics.

Organics may undergo degradation, graphitization, carbonization, and formation of SiC during thermal treatment.

5.1.3 Methods for the Determination of Surface Contaminations

In general, there are several established methods for the quantitative determination of surface impurities which were mainly developed and optimized for the microelectronics industry. We will give a short overview about those techniques and discuss the pros and cons.

Current methods for metal analyses on wafer surfaces include total reflection x-ray fluorescence (TXRF), vapor phase decomposition TXRF (VPD-TXRF), VPD inductively coupled plasma mass spectrometry (VPD-ICP-MS), and time-of-flight secondary ion mass spectrometry (ToF-SIMS). In TXRF analysis, x-rays strike the sample in a very small angle to obtain a total reflection. Thus, the penetration depth is only a few nanometers, reducing the background signal significantly. Surface contaminations can be detected due to their excited fluorescence signal emitted from the wafer surface which is detected by an energy dispersive detector [9]. This allows a quick and non-destructive analysis but is of limited sensitivity. Therefore, surface collection methods have been developed like VPD or droplet etching. In VPD, a wafer is exposed to saturated HF vapor which dissolves the oxide layer. Following, a droplet of acidic liquid is used to collect the dissolved residues from the wafer surface and to collect them into a small volume. This solution is either dried on the center of the wafer surface and analyzed by TXRF or transferred to an ICP-MS instrument. Compared to direct TXRF in VPD-TXRF, the detection limits are significantly increased. A clear drawback of the method is, however, that low Z elements like boron cannot or poorly be detected. Moreover, the drying conditions for the droplet are critical and make the procedure tedious. In contrast, VDP-ICP-MS provides the full range of elements and is sensitive also for elements like B, Na, and Al. The detection limits for both VPD methods are in the range of ~1E8–1E10 atoms/cm^2 [9]. Cl, Br, and S are not available with ICP-MS but with TXRF. A complementary technique is ToF-SIMS. It offers a wide range of application including the detection of organic compounds or particles, imaging, or depth profiling. This technique is especially surface-sensitive since it uses a pulsed ion beam (primary ions) to remove molecules from the very outermost surface of the sample. The particles are removed from atomic monolayers on the surface (secondary ions). These particles are then accelerated into a "flight tube" and their mass is determined by measuring the exact time at which they reach the detector (i.e., ToF). Essentially, it is a non-quantitative method but accurate quantification can be achieved using so-called relative sensitivity factors (RSFs) to convert normalized ion intensities to concentrations [10]. These RSFs have to be adapted for each matrix system and can be improved by calibration with a reference method like one of the above mentioned. Detection limits for alkali, earth alkali, as well as transition metals in the range of 1E7–1E10 atoms/cm^2 were reported on polished silicon wafer samples. All of the above mentioned methods were developed for the use on polished microelectronics wafer and are routinely used in semiconductor industries. The special characteristics of solar wafer, namely, the different surface quality, required the modification of those methods or the development of new techniques which are described in [11] and in the following sections.

Further methods which are basically suited for the surface analysis are laser ablation (LA-) ICP-MS and glow discharge mass spectrometry (GD-MS) since they are quantitative and require no or little sample preparation. Our own work showed, however, that both techniques are not sensitive enough for the detection of the very low level of metallic contamination usually found on solar wafers.

In contrast to the detection of metals, the analysis of organic contamination is not as advanced. Even in microelectronic manufacturing, no routine methods are used; nevertheless, a variety of techniques exists and should be introduced here: Contact angle measurement is a fast, simple, and well-known method detecting the surface hydrophobicity as an indirect measure for organic surface films. This technique can be used for the evaluation of cleaning efficiency or to characterize the influence of various environments [12] and is further described in Section 5.3.2. Absorbed organics can also be measured by infra-red absorption. Applicable techniques are MIR-FTIR or ATR-FTIR which enables the identification of different organic functions such as CH_2, CH_3, and OH. For example, IPA residues after wafer drying can be detected by MIR-FTIR [12]. Reported detection limits are in the order of 1E12 carbon atoms/cm^2 but depend on the used coupling prisms and the thickness of the air gap. Therefore, it is of limited use for rough solar wafers. As mentioned above, the ToF-SIMS technique is also suited to detect organic surface contamination. Molecular fingerprints may be used to identify chemical compounds using spectra data bases. Even though the analyzed area is very small (~300 × 300 μm^2), it could be shown that it is a useful tool for the characterization of solar wafer cleaning processes. An example is shown in Figure 5.3 [13]. Drawbacks for this method are the limited use for the recording of integral quantitative data, the rather strong influence of surface structure, and the very small detection spot. The most advanced method for the detection of organics is thermal desorption gas chromatography coupled to mass spectrometry (TD-GC-MS). Like ToF-SIMS, it can be used for polished microelectronic wafer as well as for solar wafer but is a very elaborate technique, not suited for routinely or in-line analysis. Some more details of TD-GC-MS can be found in Section 5.3.3.

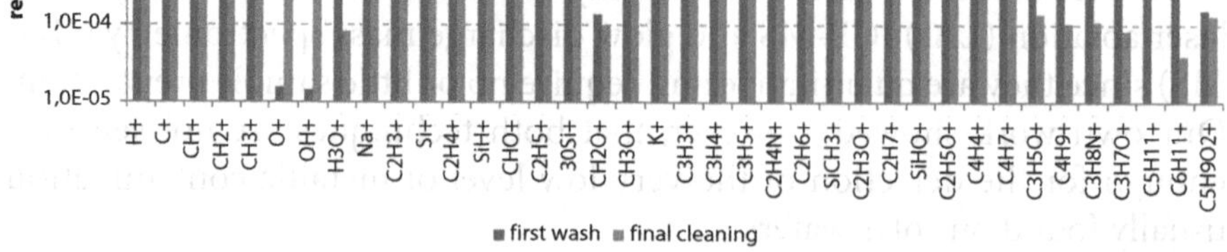

Figure 5.3 Comparison of relative signal intensities from ToF-SIMS surface spectra recorded from solar as-cut wafers after first wash and final cleaning step of the wafering process, respectively. Yellow asterisks label organic fragments which were significantly removed during the cleaning procedure. Red asterisks highlight the increased Si, O, and OH signals indicating a pure silicon oxide surface.

5.2 Metal Analysis by ICP-MS

For the surface analysis of solar silicon wafers, different methods have been developed: the sandwich method (according to [6]), the droplet scan method and the immersion method, respectively. Basically, all of the three methods could be used for wafer analysis but have different advantages and thus can be applied for different questions. The scan solution for all those methods contains diluted HNO_3 (or alternatively H_2O_2) and HF, while the first one oxidizes the silicon and the latter can dissolve the silicon oxide (Equations 5.1 to 5.3).

$$3Si + 4HNO_3 \rightarrow 3SiO_2 + 4NO + 2H_2O \tag{5.1}$$

$$SiO_2 + 6HF \rightarrow H_2SiF_6 + 2H_2O \tag{5.2}$$

$$3Si + 4HNO_3 + 18HF \rightarrow 3H_2[SiF_6] + 4NO + 8H_2O \tag{5.3}$$

Thus, not only the silicon oxide but also a thin sub-oxide layer (<2 μm) can be dissolved. According to the questioning, the scan solution can be adapted and for example HCl/HNO_3 or only HF can be used for a surface etching in order to dissolve only adhering molecules or only the silicon oxide layer.

For the analysis of one batch of identically wafers prepared with different methods of sample preparation, like sandwich and immersion method, the results correlate very well with each other. The critical factor is always the limit of detection, which is calculated from the standard deviation of the blank samples. In Figure 5.4, the limits of detection of two

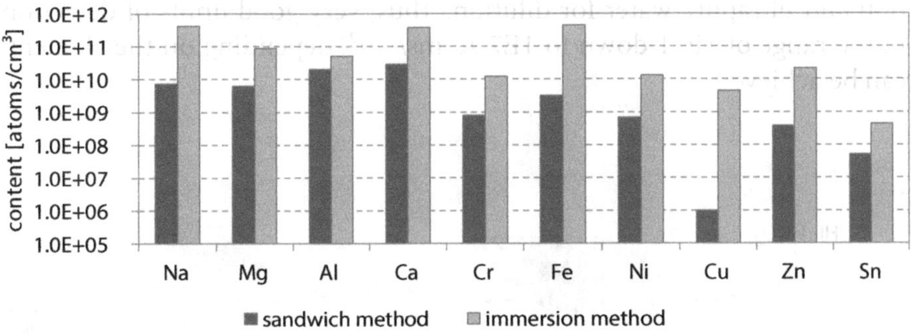

Figure 5.4 Limits of detection obtained from ICP-MS measurements comparing two methods of samples preparation. Sandwich method shows significant lower background than the immersion method and thus is much more sensitive.

different sample preparation methods and subsequent ICP-MS analysis are shown. The sandwich method shows significantly lower limits of detection compared to the immersion method and is more sensitive for measurements especially of very low contaminated samples.

For copper, some special characteristics need to be considered. Due to its noble character, there are reductive deposition processes while etching with for instance diluted HF on the hydrophobic silicon wafer. Thus, high concentrated H_2O_2 scan solutions are used to keep the copper soluble and make sure that the collecting efficiencies are higher [14].

Also, it is known that some elements, again especially copper, are very diffusible in the silicon matrix. Experiments with spiked wafers, and subsequent determination of the recovery rates by etching with HF/HNO_3, have shown that copper can be recovered from the wafer surface only to a very low percentage. Further investigations brought the result that copper diffused into the bulk material [15]. Therefore, whenever establishing a new method, recovery rates have to be determined carefully.

5.2.1 Sandwich Method

The recommended method for surface extraction is the sandwich method [6] for which a pair of identically wafers is needed. One milliliter of etching solution is applied to one wafer surface (volume appropriate to a standard 6-inch wafer) in small droplets and then covered with a second identical wafer. It is best to use neighbored wafers of the same batch. After 10 minutes, the two wafers can be separated, the droplets of etching solution are collected using a pipette, and the solution is measured by ICP-MS after dilution with ultrapure water (Figure 5.5). Blank samples to determine the background of the method contain only the same amounts of etching solution and ultrapure water for dilution. Thus, very good limits of detection in the range of 1E11 down to 1E7 atoms/cm^2, depending on the element, can be achieved.

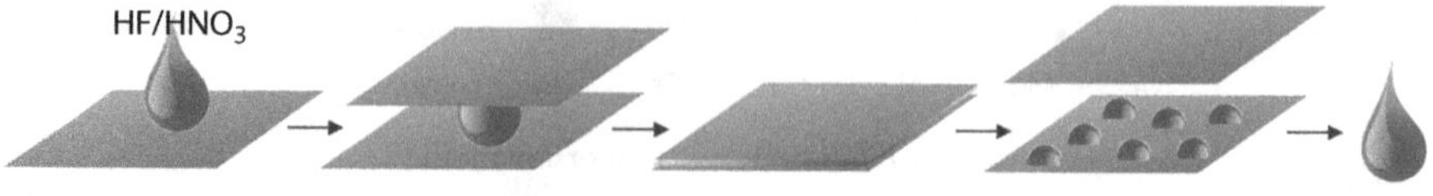

Figure 5.5 Schematic representation of the sandwich method: etching solution is applied between two wafer surfaces, contaminations are extracted, and the final solution can be measured by ICP-MS after dilution.

This method is only practicable for wafers with hydrophobic surface, because only then the droplets can be collected after the extraction time. If the surface is too hydrophilic, the etching solution will be spread as a liquid film over the whole surface and only a small percentage can be collected after the extraction time. Thus, results would be erroneous due to extrapolation to 100% of etching solution. For those wafer types, the immersion method (Section 5.2.3) is more suitable.

The big advantage of the sandwich method is that two wafer surfaces are extracted and combined in one sample solution which improves the sensitivity of the method by the factor of two.

5.2.2 Droplet Scan Method

The droplet scan method is a modification of the VPD method which is not applicable for as-cut solar wafers. It is a preconcentration method which was originally used to improve the limits of detection for the TXRF measurements but can also be used for subsequent ICP-MS analysis.

For the automated VPD procedure, the silicon wafer is exposed to HF vapor, where the thermal, chemical, or native silicon oxide is etched according to Equation 5.2. All contaminations, which are located on top or inside the oxide layer, can be collected by a following scanning procedure with about 100 μl of a scan solution (volumes can deviate between different laboratories) which contains HF/HNO_3 or HF/H_2O_2. The concentration of the scan solution can vary between different laboratories but is mostly in the lower percentage range. The scan solution can finally be diluted and analyzed by ICP-MS (Figure 5.6). The scanning procedure is possible due to the hydrophobic surface properties after the exposition step in HF vapor [14].

As the VPD method is not suitable for as-cut solar silicon wafers, the method was adapted to the requirements of the rough as-cut surfaces. Basically, a droplet of scan solution (containing HF/HNO_3) is lead over the whole surface of a wafer collecting all surface impurities in a very small amount of etching solution without prior exposition in HF. For the manual handling a droplet volume of about 0.25 ml of etching solution will

Figure 5.6 Schematic representation of the VPD method. After an exposition in HF vapor, the surface contaminations can be collected by a scan solution and finally be measured by ICP-MS.

be balanced over the whole wafer surface and afterwards collected by a microliter pipette. With a second sampling step with 0.25 ml of fresh etching solution, it can be ensured that the whole surface is scanned and all impurities are collected in the etching solution which can be analyzed for their elemental contents by ICP-MS afterwards. Blank samples to determine the background of the method contain only 0.5 ml of etching solution and the used amount of ultrapure water for dilution. Compared to the sandwich method, the sensitivity is by factor two lower as only one wafer surface (instead of two as for the sandwich method) is analyzed. Thus, the limits of detection are again in the range of 1E11 to 1E7 atoms/cm^2, depending on the element. The droplet scan method is suitable for wafers with hydrophobic surfaces but causes some problems on hydrophilic surface due to spreading of the droplet.

5.2.3 Immersion Method

For wafers with a very hydrophilic surface, neither sandwich extraction nor droplet scan is suitable due to high spreading of the etching solution. Thus, the wafer surfaces can be extracted in a bag of inert polymer foil (or appropriate PFA vessels) with the etching solution. The bags need a special cleaning procedure prior to further usage to ensure a low background. Rinsing should be proceeded with etching solution and ultrapure water several times before the analysis takes place and also a measurement of the last rinsing solution (best: etching solution) should be performed to ensure a low background of contaminations. The surface extraction for a 6-inch wafer requires a higher volume of etching solution to cover the whole sample from both sides, etching times can vary between different laboratories, but should be at minimum 10 minutes to ensure a complete etching and dissolution. During analysis, at least two bags with no sample should be carried along to determine the influence of the solution, the polymer bags, and the handling itself (Figure 5.7).

The whole procedure is more time-consuming due to plenty of handling steps and the sensitivity is lower. Especially, the total amount of impurities

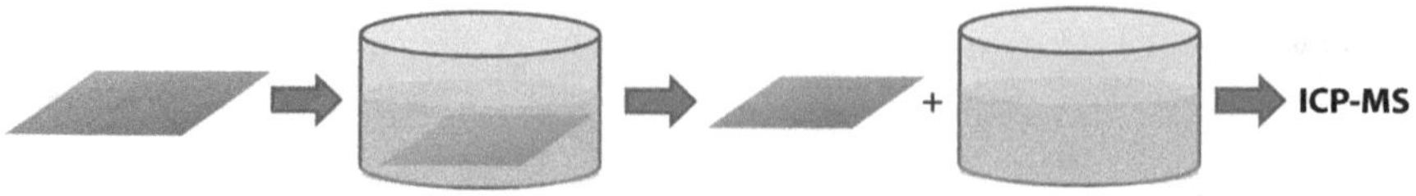

Figure 5.7 Schematic representation of the immersion method: a wafer is extracted in etching solution in a bag of inert polymer foil and the resulting solution can be analyzed by ICP-MS.

removed from the wafer surfaces is diluted in a significant higher volume, and thus, sensitivity of the method is reduced. The method should be only used for wafers where no other method can be performed or very high contamination levels are expected.

5.2.4 Sequential Etching Methods

For the detection of trace elements which may be located in different layers of the material, appropriate etching methods are available. Due to the variation of etching solutions and their chemical reaction behavior elemental contents on the surface, like adhesive particles, included in the silicon oxide and sub-oxide layers as well as the silicon bulk material can be analyzed. The method is available for all types of samples, like wafers, powders, granulate, or chunks and the handling can be adapted according to the shape of the sample [16].

For silicon wafers the methods described in Section 5.2 are available for appropriate analysis. Depending on the hydrophobic character of the sample the sandwich method is again recommended due to best limits of detections. Alternatively, especially when higher concentrations of etching solutions are necessary to remove thicker layers, an immersion method in PFA vessels is of advantage. If the method is adopted for silicon chunks or powder samples, inert vessels should be used where the etching solution is filled in until fully coverage of the sample piece. For all sample types, the determination of multiple samples is recommended to get statistics of the material and to identify outlier values.

Elemental contaminations which are attached to the surface as adhesive particles can be dissolved by diluted HNO_3/HCl in ultrapure water (volume ratio 1:1:23). As no HF is in the etching solution, the silicon itself is not affected and a pure surface analysis can be performed. For the next etching step diluted HF in water (volume ratio 1:24) will be used and thus only the few μm thick silicon oxide can be dissolved according to Equation 5.2. That step is followed by etching with HNO_3/HF where the silicon is oxidized by the HNO_3 and the silicon oxide dissolved by the HF (according to Equations 5.1 to 5.3). Depending on the concentration of the acids as well as the etching time, the depth of the removed layer can be varied. Higher concentrations as well as longer etching durations will result in higher etch removal. Appropriate vessels, like PFA or PTFE vessels, should be used with higher concentration due to higher vaporization and thus losses of etching solution and due to higher safety risks. It is also possible to repeat that etching step several times to receive a rough depth profile of contamination. At the end, a digestion of the residual sample can be performed to determine the bulk contamination of the sample. The whole procedure using different etching steps is shown schematically in Figure 5.8 [16].

In parallel to all the single etching steps, a complete digestion of the whole sample should be performed. The sum of the single etching steps should correlate to the total amount of impurities from the one-step decomposition. This can be used as an internal control whether the whole analysis was carried out correctly. Small deviations can occur due to the usual failure of analysis, but if higher deviations occur, a critical review of the process should be done. Figure 5.9 shows exemplarily the results for the whole analysis in four steps of a metallurgical silicon powder. The correlation between the single etching steps and the one-step decomposition is shown in Figure 5.10.

Due to the results achieved from the consecutive etching steps, a localization of impurities in different layers, and thus, a correlation to their origin can be determined.

While bulk contaminations are more likely to be introduced in the crystallization process, contaminations within the silicon oxide layer may be introduced by the wafer sawing process where silicon bonds were separated and therefore the silicon is bare for the first time. While the native silicon oxide layer is growing, contaminations from the process could be included in the molecule grid structure. Adhesive particles can attach to

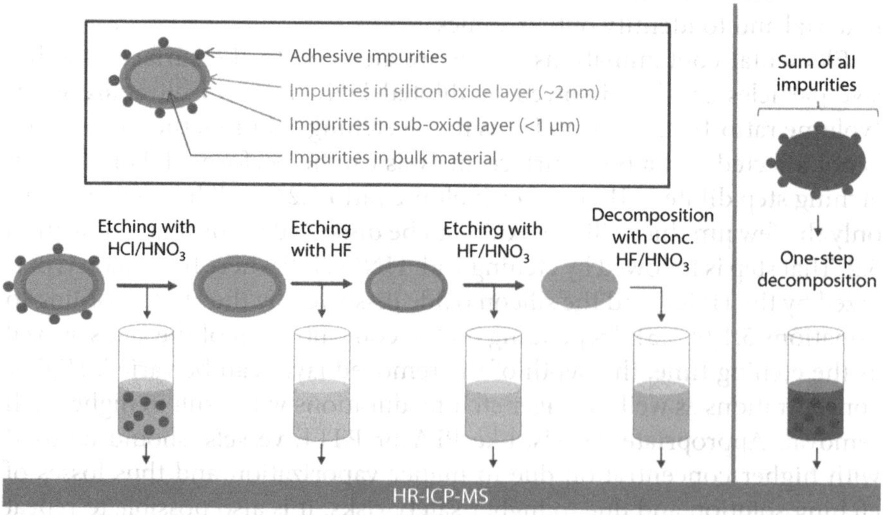

Figure 5.8 Schematic representation of the sequential etching method to locate impurities in different layers of the appropriate sample.

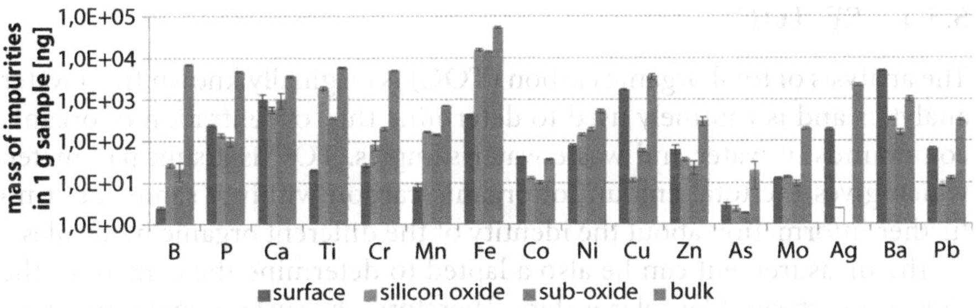

Figure 5.9 Elemental contaminations in different layers of a metallurgical silicon powder. Standard deviations are refered to multiple measurements and colorless bars show values below the limit of detection.

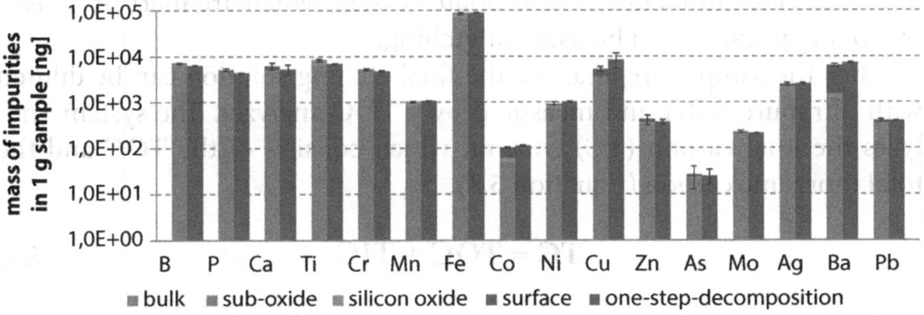

Figure 5.10 Summed up masses of impurities of sequential etching steps and final decomposition of the samples (stacked bars) compared to the amount of impurities after one-step decomposition (red bars).

the surface due to all the following process steps but should be removed in the whole cleaning process which is performed straight after wafer sawing.

5.3 Determination of Organic Contaminations

The analysis of organic impurities is important, as they have detrimental impacts on further process steps as shown in Section 5.1. This section introduces three methods to detect organic impurities. On the one hand, integral analysis can be performed by surface extraction total organic carbon (SE-TOC) measurement giving an integral value over the whole wafer surface and, on the other hand, by determination of the surface energy providing lateral information of the contamination on wafer surfaces. As a further state-of-the-art method, TD-GC-MS analysis is described at the end of the chapter.

5.3.1 SE-TOC

The analysis of total organic carbon (TOC) is originally known from water analytics and is routinely used to determine the concentration of organic compounds in water and waste water samples. TOC is a sum parameter, which gives the total amount of organic carbon within a sample, but no further information about the identity of the different organic molecules.

The measurement can be also adapted to determine the content of the TOC on surfaces, like solar wafers. Therefore, the surface contaminations have to be extracted from the wafer surface by one of the methods shown in Section 5.2. Best sample preparation can again be achieved by the sandwich method or the droplet scan method. Also the immersion method would be possible, but should not be performed in polymer foil bags due to organic contamination. Appropriate vessels, which are inert or at least free of organics, should be used for etching.

After the sample preparation, the final etching solution can be diluted with ultrapure water and measured by a TOC analyzer. The system analyzes the total carbon (TC) content, which consists of the TOC and the total inorganic carbon (Equation 5.4).

$$\mathbf{TC = TOC + TIC} \tag{5.4}$$

The TOC consists of non-purgeable organic carbon (NPOC) and purgeable organic carbon (POC, Equation 5.5). While the first one remains in the sample after purging with nitrogen, the latter will be removed due to its low vapor pressure. Thus volatile organic compounds, which are not expected to have detrimental impacts on silicon wafer surfaces, will be removed and only the NPOC content will be measured.

$$\mathbf{TOC = NPOC + POC} \tag{5.5}$$

The carbon in organic compounds will be converted into carbon dioxide (CO_2) by oxidants, UV, or higher energetic light. Persulfates are used as oxidant agents and the oxygen-oxygen bonds are split by UV light into sulfate radicals (Equation 5.6).

$$\mathbf{S_2O_8{}^{2-} \xrightarrow{E_{UV}} 2\ SO_4{}^{2-}} \tag{5.6}$$

A chain reaction is implicated which leads to the complete oxidization of the organic compounds (Equation 5.7). Water will be filtered and the carbon dioxide content can be measured by a CO_2 selective NDIR detector.

$$R{-}H \xrightarrow{[O],\ E_{UV}} CO_2 + H_2O \tag{5.7}$$

For samples which may contain inorganic carbon, they will be acidified prior analysis. Thus, the chemical equilibrium will be shifted to CO_2 (Equation 5.8) and can be removed by purging with nitrogen.

$$CO_2 + H_2O \leftrightarrow H_2CO_3 \leftrightarrow HCO_3^- + H^+ \leftrightarrow CO_3^{2-} + 2H^+ \tag{5.8}$$

After the whole procedure of acidifying, purging, and oxidizing, only the content of TOC is measured. A value for the organic carbon of the wafer surface can be calculated, and thus, an integral method for the determination of organic carbon on solar wafer surfaces is available. Limits of detection for the standard 6-inch solar wafer are at 1E13 atoms/cm^2 and therefore completely sufficient for the determination even for low contaminated samples. Results of wafers from different cleaning steps are exemplarily shown in Figure 5.13. Although the chemical identity of the organic molecule is still unknown and the origin of the contamination cannot be assigned, the TOC value gives an impression of the amount of organic compounds and hence can be used to evaluate cleaning strategies [17].

5.3.2 Contact Angle Mapping

By the measurement of contact angles on the surface of silicon wafers with lateral resolution, information can be gained about the surface distribution of organic contamination. If the contact angles are determined with different liquids like water and diiodomethane, the surface energy may be calculated.

Performing the described method a pattern consisting of 30 equidistant distributed positions (for standard 6-inch wafers) is measured. At each point a droplet of water is placed on the wafer surface by a vertical syringe, the contact angle Θ is measured automatically by the software of the measuring system and afterwards the droplet is suctioned by another syringe to remove it. The procedure is repeated with diiodomethane just next to the place where the water droplet was located. The whole wafer is scanned dropled by dropled with that method. The suction step after each droplet is important; otherwise, the droplets will overlap in the camera view and no useful results could be gained.

The contact angle is the angle which is formed between the liquid-solid and the liquid-vapour interface. The angles size depends on the interaction between the substances at the points of contact. The lower the interaction is, the higher is the contact angle, which can be measured by the static sessile drop method. A tangent will be applied to the droplets profile in the area of the three-phase point and the contact is determined at both sides of the droplets image (Figure 5.11). The average value of both contact angles will be used for further calculations. The reproducability is at its best at ± 0.2°. The whole procedure can be performed automatically by a software.

From the contact angles, measured with water and diiodomethane, the surface energy can be calculated according to the formula of Owens and Wendt [18] (Figure 5.12). The surface tension values for polar proportion of 46.8 and 1.3 mN/m and dispersive portion of 26.0 and 49.5 mN/m for water and diiodomethane, respectively, will be used.

A graphic representation helps to get an impression of the lateral distribution of surface energies on the wafer surface and to draw conclusions on the distribution of organic contaminations. Areas with lower organic contaminations show lower surface energies, while highly contaminated areas have much higher surface energy values.

In Figure 5.13, a comparison of data obtained from SE-TOC and contact angle measurements is shown for wafer samples from different process

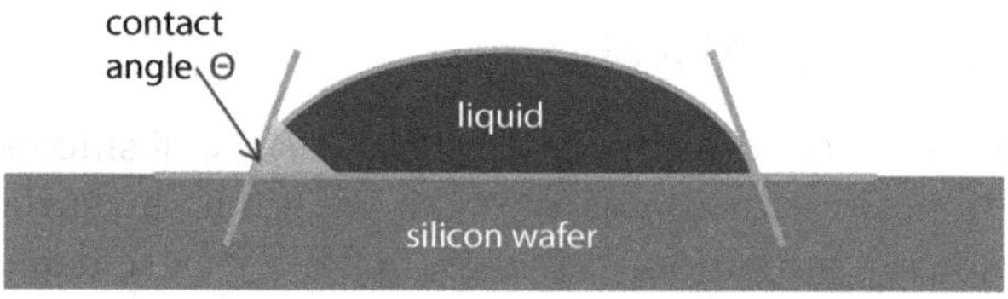

Figure 5.11 Measured contact angle of a liquid on silicon wafer surface at the three-phase point between the silicon wafer, the liquid, and the surrounding air.

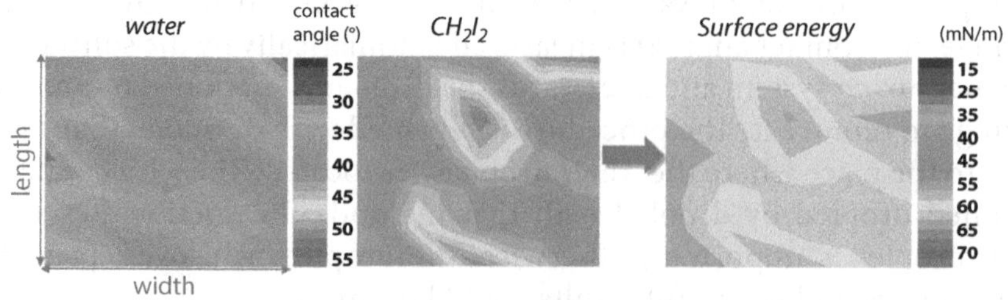

Figure 5.12 Wafer map of the contact angles measured with water and diiodomethane and the calculated surface energy map.

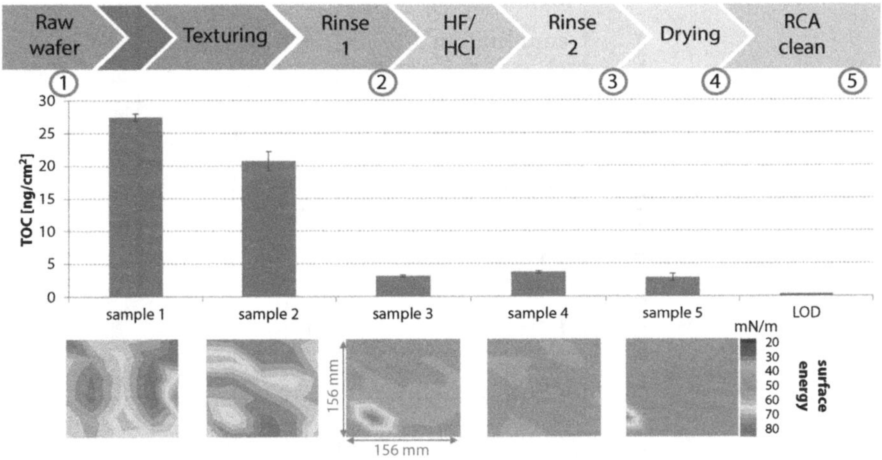

Figure 5.13 Results for TOC measurements and from contact angle calculated surface energy maps of silicon wafers from different process steps shown.

steps. A clear correlation of both methods can be observed in that a significant increase of surface purity is consistently detected after the HF/HCl dip [19].

While the TOC measurement can provide a quantitative value of the TOC, the surface energy can show the distribution of the organic contamination on the wafer surfaces. Even though neither of the two methods leads to more knowledge about the chemical structures of the organic molecules, both methods provide valuable information about the organic contaminations of the silicon wafers and can be used to evaluate the wafer cleaning procedures.

5.3.3 TD-GC-MS

The most advanced technique for the detection of organic surface impurities is TD-GC-MS. It combines the superior separation ability of GC and the powerful unknown identification and quantification ability of mass spectrometry. The TD-GC-MS technique allows qualitative and quantitative analysis of volatile and semi-volatile compounds outgassing from the wafer surface during a certain temperature ramp [20]. The clear advantage of this method is that the adhering molecules can be identified and this may help to find the source of contamination. An example is given in Table 5.1, where wafers from a batch showing significant texture defects were analyzed by TD-GC-MS in comparison to an unstained reference batch.

Table 5.1 Semi-quantitative results of a TD-GC-MS analysis of organic surface contamination on solar as-cut wafers. The last raw gives the sum of all detected signals, including also unknown compounds.

Compound name	**methyl dioxolane**	**1,4-dioxane**	**styrene**	**2,2'-oxybis-ethanol**	**benzaldehyde**	**bencoic acid**	**ketone**	**1-dodecanol**	**1-tridecene**	**1-pentadecene**	**hexadecanol**		**sum [ng/cm^2]**
Reference wafer	++	+	+	-	+	-	-	+	++	++	++		**4**
Wafer with optical defects after texturing	+	+	++	++	++	++	++	+	++	-	-		**25**

Of course, significant drawbacks of this technique are the high complexity of instrumental equipment necessary and rather time consuming procedure.

Acknowledgments

The authors would like to thank Dr. Susanne Richter (Fraunhofer Center for Silicon Photovoltaics CSP) for information about ToF-SIMS and providing Figure 5.3.

References

1. Kashkoush, I., Chen, G., Nemeth, D., Rieker, J., Wafer surface preparation for high-efficiency solar cells. *40th IEEE Photovoltaic Specialists Conference*, pp. 1211–1213, 2014.
2. Kranz, C., Wyczanowski, S., Baumann, U., Dorn, S., Queisser, S., Schweckendiek, J., Pysch, D., Dullweber, T., Industrial Cleaning Sequences for Al 2 O 3-passivated PERC Solar Cells. *Energy Procedia*, 55, 211–218, 2014.
3. Meißner, D., Meyer, S., Anspach, O., Enrichment of metal ions in virgin Si-surfaces. *Energy Procedia*, 27, 27–32, 2012.
4. Buchholz, F. and Wefringhaus, E., Impact of Fe and Cu surface contamination on high efficiency solar cell processes. *Sol. State Phenom.*, 219, 305–311, 2015.
5. Al-Amin, M. and Assi, A., Metal oxides for photo-induced hydrogen production and dye-sensitized solar cell applications. *Materials and Processes for Energy: Communicating Current Research and Technological Developments*, First edition, A. Méndez-Vilas (Ed.), pp. 867–879, Formatex Research Center.
6. Buchholz, F., Wefringhaus, E., Eisert, S., Egenolf, H., Groß, A., Measurement and impact of surface transition metal contamination of textured multicrystalline silicon wafers. *26th European Photovoltaic Solar Energy Conference and Exhibition*, pp. 1187–1190, 2011.
7. Sonner, C., Oltersdorf, A., Zimmer, M., Rentsch, J., Influence of contaminations and cleaning sequences on alkaline texturisation. *26th European Photovoltaic Solar Energy Conference and Exhibition*, 2011.
8. Choi, K., Ghosh, S., Lim, J., Lee, C.M., Removal efficiency of organic contaminants on Si wafer by dry cleaning using UV/O 3 and ECR plasma. *Appl. Surf. Sci.*, 206, 1, 355–364, 2003.
9. Wang, J., Balazs, M.K., Pianetta, P., Baur, K., Brennan, S., Analytical techniques for trace elemental analyses on wafer surfaces for monitoring and

controlling contamination. *19th Annual Semiconductor Pure Water and Chemicals Conference*, 2000.

10. Rostam-Khani, P., Hopstaken, M.J., Vullings, P., Noij, G., O'Halloran, O., Claassen, W., Quantification issues of trace metal contaminants on silicon wafers by means of TOF-SIMS, ICP-MS, and TXRF. *Appl. Surf. Sci.*, 231, 720–724, 2004.
11. Oltersdorf, A., *Spurenanalytische Untersuchung der nasschemischen Prozesse in der Siliziumphotovoltaik*, PhD Thesis University Freiburg, Freiburg, 2014.
12. Roche, A., Wyon, C., Marthon, S., Ple, J.F., Olivier, M., Rochat, N., Chabli, A., Danel, A., Juhel, M., Tardif, F., Detection of organic contamination on silicon substrates: Comparison of several techniques. *Characterization and Metrology for ULSI Technology 2000*, 2001.
13. Meyer, S., Wahl, S., Timmel, S., Richter, S., Gebhardt, S., Characterization of Wafer Surfaces: Analytics for sawing and purification processes. *Workshop Elemental and Surface Analytics in Photovoltaics*, Halle (Saale), 2012.
14. Metz, S., *Analytik von Kontaminationen auf Siliciumoberflächen: Möglichkeiten und Grenzen des VPD-Verfahrens*, PhD Thesis University Frankfurt am Main, p. 6, 2004.
15. Meyer, S., Wahl, S., Hagendorf, C., Advanced Metal Contamination Analysis for High Efficiency Solar Cell Manufacturing. *Energy Procedia*, 92, 369–373, 2016.
16. Wahl, S., Meyer, S., Hagendorf, C., Localization of Inorganic Impurities in Silicon Samples by Sequential Etching and ICP-MS Detection. *Energy Procedia*, 92, 392–398, 2016.
17. Meyer, S., Wahl, S., Timmel, S., Köpge, R., Jang, B.-Y., The impact of wafering on organic and inorganic surface contaminations. *Appl. Surf. Sci.*, 378, 384–387, 2016.
18. Owens, D.K. and Wendt, R.C., Estimation of the Surface Free Energy of Polymers. *J. Appl. Polym. Sci.*, 13, 8, 1741–1747, 1969.
19. Meyer, S., Timmel, S., Hagendorf, C., Rapid Determination of Organic Contaminations on Wafer Surfaces. *Sol. State Phenom.*, 219, 317–319, 2014.
20. Sun, P., Ayre, C., Wallace, M., Characterization of organic contaminants outgassed from materials used in semiconductor fabs/processing. *AIP Conference Proceedings*, vol. 683, pp. 245–253, 2003.

Index

Also of Interest

Check out these related books on Photovoltaics

Photovoltaic Manufacturing
Texturing, Etching, and Cleaning
Edited by Monika Freunek Müller
Published 2021. ISBN 978-1-119-24189-8

Indoor Photovoltaics
Edited by Monika Freunek Müller
Published 2020. ISBN 978-1-119-60559-1

Green Energy
Solar Energy, Photovoltaics, and Smart Cities
Edited by Suman Lata Tripathi and Sanjeevikumar Padmanaban
Published 2020. ISBN 978-1-119-76076-4

Photoenergy and Thin Film Materials
Edited by Xiao-Yu Yang
Published 2019. ISBN 978-1-119-58046-1

Photoelectrochemical Solar Cells
Edited by Nurdan Demirci Sankır and Mehmet Sankır
Published 2019. ISBN 978-1-119-45993-4

Emerging Photovoltaic Materials
Edited by Santosh K. Kurinec
Published 2019. ISBN 978-1-119-40754-6

Photovoltaic Modeling Handbook
Edited by Monika Freunek Müller
Published 2018. ISBN 978-1-119-36352-1

Printable Solar Cells
Edited by Nurdan Demirci Sankir and Mehmet Sankir
Published 2017. ISBN 978-1-119-28371-3

Handbook of Cleaning for Semiconductor Manufacturing
Fundamentals and Applications
Edited by Karen Reinhardt
Published 2011. ISBN 978-0-470-62595-8

Printed in the USA/Agawam, MA
September 16, 2026

910661.003